I0119213

LET'S TRADE, NOT ARGUE

First Published in 2025 by Echo Books

Echo Books is an imprint of Superscript Publishing Pty Ltd,
ABN 76 644 812 395
Registered Office: Post Office Box 669, Woodend, Victoria, 3442
www.echobooks.com.au

Copyright © 2025 Bob Breen and Ian Langford

ISBN: 978-1-922603 978-1-922603-76-0 (paperback)

LET'S TRADE, NOT ARGUE

An Australian strategy to secure a respectful relationship with China

Bob Breen and Ian Langford

echo))

BOOKS

DEDICATION

Bob Breen dedicates this book to his patient and supportive wife, Rhonda, Associate Professor Mitchell Hansen and his team for their life-saving surgery at 3 am on Saturday, 13 July 2024, at John Hunter Hospital, Newcastle, after Bob's emergency evacuation by helicopter from Coffs Harbour Hospital.

Ian Langford dedicates this book to Warrant Officer Class One RJ Langford, Vietnam War veteran, distinguished soldier and quiet Australian, and Deakin Distinguished Professor Matthew Clarke in gratitude for loyal support.

ACKNOWLEDGEMENTS

Professor Clive Hamilton's books, Dr Ross Babbage's Center for Strategic and Budgetary Assessments monographs, and the work of others, such as Emeritus Professor John Fitzgerald, Professor Mary-Anne Brady, Dr Alan Dupont, Peter Hartcher, and Dr David Kilcullen, have significantly influenced the content of this book. Their invaluable contributions and insights have been instrumental in shaping the proposed strategy and Response Force concepts against the danger. The authors acknowledge the book's anonymous 'insider' reality checkers. Their work forms a crucial part of this book's intellectual journey.

Emeritus Professor David Horner's histories of Australia's preparations for and participation in the Pacific War and the development of the Australian intelligence services and special forces informed the authors' proposals for establishing a unique Australian Response Force for national resilience and deterrence in the grey zone.

The authors acknowledge their colleague, Dr Mark Armstrong, for his PhD thesis, proving the need to transform voluntary Australian military service into voluntary national service for national resilience and grey zone defence.

PREFACE

The CCP is 'on the march' in the grey zone, a nation's legal and illegal actions conducted between the 'white' of peace and the 'black' of war under the threshold of acts of war to dominate another nation or seize territory. This threat is not a distant possibility but a present reality, demanding immediate attention and action. The CCP aims to persuade and, if that doesn't work, to coerce Australia and its neighbours to become China's Pacific region tribute states.

There is no geographic frontline. Unlike Japanese nationalists who occupied northeast and southeast Asia after military conquest, the CCP targets the Australian economy, political system, public opinion, essential services, financial systems, and supply chains to coerce Australia into a relationship on its terms.

The CCP has a three-phase light-to-dark grey zone campaign plan. The Chinese President, Xi Jinping, launched the 'light-grey' phase in 2014 with expressions of goodwill and a free trade agreement. By 2021, Australia had reasserted its US alliance. CCP grey zone actions escalated to 'darker grey' aggressive diplomacy, espionage, political interference, cyber-attacks, trade embargos and the infamous dossier of 14 CCP grievances against Australia. That's how these campaigns escalate and oscillate.

The CCP, frustrated with the Abbott, Turnbull, and Morrison governments, paused its bullying campaign in 2022 and waited for the next election. At the time of writing, in November 2024, the CCP is courting rather than coercing. China has eased trade embargos. The Albanese government and CCP have had meetings, visits, and negotiations, signed trade and other agreements, and announced that the relationship is 'back on track' at the G20 Summit in Rio De Janeiro.

In 2023, the CCP also paused the more abrasive side of its worldwide campaign, waiting for the US presidential election results. Now that President Donald Trump has returned for a second term, the CCP will negotiate with him about Taiwan and 'press to test' elsewhere. A deals-based world order will replace a rules-based global order. Australia and other US allies in the Asia-Pacific region will be 'on the table'. President Trump may do an 'America First' deal on debt, trade and spheres of influence. If the United States remains loyal to its allies, the CCP will escalate in the grey zone but not go to war.

There will be two possible triggers for CCP grey zone escalation against Australia in 2025/26. A re-elected Albanese government or first-term Dutton government will affirm Australia's US and UK alliance, as exemplified by the AUKUS agreement (2021 agreement for defence collaboration between Australia, the US and the UK). Concurrently, President Trump may compromise on Taiwan to avoid a Third World War. If he chooses not to compromise or to continue with diplomatic ambiguity, the CCP will likely escalate in the grey zone to achieve its strategic objectives without war. A deal between China and the United States to avoid a Third World War will leave Australia alone and unprepared for a grey zone escalation.

Much has been written about the CCP grey zone threat, but no one has shown a way to stop it. This book does. Readers should examine the motives of those who criticise the authors and this book and ask them if they have a better plan because no sensible person denies the threat.

This book proposes a complementary strategy for managing Australia's relationship with China. It seeks one based on trade, not one complicated by CCP rivalry with the United States and aspirations to dominate the Asia Pacific region. It deters the CCP from escalating against Australia by messaging 'Let's Trade, not Argue'. The challenge is creating a fourth armed service, a Response Force, to deter the CCP from ignoring the invitation and to defend Australia against the CCP in the grey zone with sovereign resources.

The book is a 'wake-up call' to spark public opinion by proving that a CCP grey zone campaign can and will endanger Australians and their way of life if Australia does have a de-escalation strategy and a Response Force. It seeks public support for a plan that no individual department or government agency has the will or remit to implement. Ultimately, the Prime Minister and his department must lead in marshalling political, government, corporate, and public support. The Prime Minister and his government are unlikely to read this book and act. But he and his advisers and colleagues will act on public opinion.

After President Trump takes office in 2025, the CCP will 'press to test'. Australia's second honeymoon with China will end. By then, Australia must know what it is doing in the grey zone and be ready to thwart escalation. Xi Jinping turned 71 in June 2024. He is not waiting much longer to change the world. The crucial question for the Prime Minister, his government, corporate Australia and all Australians is, 'How can I contribute to getting Australia ready in the grey zone?'

Bob Breen and Ian Langford
November 2024

About the authors and the book's proposals

The Powerful Owl Program

An innovative academic program brought the authors together. In 2016, the Department of Defence Education and Training Board endorsed Deakin University's proposal to offer senior Defence officers and public servants a professional PhD program. Most Australian universities, including ANU and UNSW, offer doctorates in various professions. Deakin University was the first to offer one in strategic studies for Defence professionals. Until the University discontinued the program in 2022, it facilitated specialised, deep Defence-relevant academic research and exposition. Candidates received recognition for their professional practice and writing, achieved higher intellectual mastery at PhD level and contributed to the theory and practice of defending Australia and its national interests.

The program attracted an eclectic candidate group who adopted the Australian Powerful Owl to represent their aspirations to combine career wisdom and research excellence to develop new 21st-century Australian strategic thinking. This bird is Australia's largest and most intelligent owl, with acute hearing and eyesight, especially at night. It is an apex predator and hunter who signals its home territory with a slow, ominously uttered 'wooo hooo' warning call. It flies low and hard at long ranges to protect its territory and kills surprised prey ten times its body weight.

The book argues for an Australian middle power strategy to emulate Powerful Owl attributes: astute situational awareness, stealth, and the smarts to detect, stare, warn and deter, and if deterrence does not work, to de-escalate. If de-escalation calibrated to prompt a return to negotiation and a fair trading relationship does not work, stealthy and lethal attributes that magnify persuasiveness by ten will be decisive.

Credible authors?

Professor Bob Breen OAM is a published soldier-scholar with over 40 years of military and academic experience. He is a former Army Colonel and a veteran of Australian peace support operations in the 1990s and 2000s, serving the Australian government on contract in Afghanistan from 2012 to 14.

Bob directed the Defence Professional PhD program until his retirement in 2022 and appointment as an Honorary Professor at the Australian War College in 2023. His PhD thesis critiqued Australian military force projection. In 2008, he published *Struggling for Self-Reliance: Four Case Studies of Australian Regional Force Projection in the late 1980s and the 1990s*. He self-published *Disappointing the Dragon: How Australia Should Stand Up to the Communist Party of China* in 2022 to warn of the Chinese grey zone threat and Australia's vulnerability and propose a de-escalation strategy.

Professor Ian Langford, DSC and Bars, is a lecturer at UNSW and Executive Director of S&D PLuS, a consortium of Arizona State University, King's College London, and the University of New South Wales to solve global challenges.

Ian is a former Army Brigadier and a veteran of Australia's special forces who participated in the War on Terror in the 2000s and 2010s. He has also officially served the ADF in China, observing and training alongside PLA forces. His PhD thesis discussed new modes of special warfare and proposed enhancing Australia's special warfare capabilities to prosecute a de-escalation strategy in the grey zone.

Credible ideas?

The book is the confluence of Bob Breen's ideas and those of Ian Langford and Mark Armstrong, whom he supervised for their professional PhDs by Folio. Mark Armstrong, CSC, is the Queensland State Emergency Service Chief Officer and Army Reserve Brigadier who served as the Adjutant General at Army Headquarters in 2024. He is a veteran of Australian homeland response operations in the aftermath of floods and the COVID-19 pandemic in 2020–21. His PhD was about mobilising Australia's human capital through innovative part-time military service for national disaster resilience and grey zone defence – a modern Army Reserve for the 21st Century.

The book draws on Mark Armstrong's PhD thesis. Still, he is not a co-author because it extends his research to a critique of Defence's stewardship of the Army Reserve and offers recommendations for substantial change to the status quo that should not be attributed to him as a co-author. Breen and Langford make recommendations for change that Armstrong believes are worthy of consideration, but they are not his conclusions or recommendations.

Disappointing the Dragon extended Langford's thinking. It added a first phase (Phase 1) that paced the escalating non-military and non-violent lighter shades of grey with deterrent diplomacy backed by a Response Force that could 'sting' and 'jab' to get attention. This book adds a second defence phase (Phase 2) that employs the Response Force as a fourth armed service to pace and de-escalate increasingly assertive, violent and destructive grey zone tactics and create deterrence while messaging 'Let's trade, not argue.' For this book, Langford's PhD proposal becomes the third defence phase to pace and de-escalate armed and violent dark shades of grey (Phase 3) with a Response Force component able to conduct extraordinary operations to punch harder for specific strategic effects anywhere and anytime.

An Australian Response Force is new. Its members are the best of Australia's human capital, employed full-time and part-time in the grey zone, which creates uncomfortable consequences for ignoring Australia's invitation to trade, not argue. In the first and second phases, its units conduct offensive and defensive, primarily informational and cyber operations, to create deterrence and de-escalate. In the third phase, an armed component of the Response Force pre-empts hybrid warfare forcefully. It has special forces attributes but is not 'of' the ADF's special forces or a branch of the ADF's special forces. Its attributes are informed by Australia's unfulfilled special operations of the Pacific War, not any other nation's armed intelligence service. It conducts extraordinary operations to protect Australia – Australia's Powerful Owl ultimate deterrent.

The book revives the core idea of part-time military service, inspiring every capable citizen to contribute to Australia's national security, not just for territorial defence from invasion. It emphasises national resilience that accords with the aspirations of the 2024 National Defence Strategy (NDS2024) 'to anticipate, prevent, absorb and recover from natural and human-induced threats and hazards' ... and 'enable more effective responses to foreign interference, espionage, terrorism and violent extremism.'

Armstrong's PhD thesis complements the book's proposed Response Force three-phase grey zone defence with a new vision for Australian part-time military service for national resilience from natural and grey zone-made disasters. The book extends his thesis's conclusions by proposing transforming Australia's current voluntary armed part-time military service arrangements, the Army Reserve, into voluntary unarmed and armed part-time national service to defend Australian lives and property from natural disasters and the disaster-like effects of cyberattacks on supply chains and essential services that can displace and distress communities. The Response Force Reserve will be a more resilient and robust voluntary national service to defend the homeland, the original vision for part-time military service from colonial times.

NDS2024 focuses the ADF on the air-sea gap around and to the north of Australia to defeat an invasion fleet and its air and missile cover. In contrast, the book's proposed Australian full-time and part-time Response Force counters the CCP grey zone campaign in the homeland and will sting and jab overseas to send messages and deter escalation. It is a war prevention strategy, but if it fails, the Response Force will project force to meet and disrupt an invasion fleet and its air cover in home ports and airfields.

Evidence?

Where is the evidence to support prophesising grey zone escalation, doomsday warnings and calls for a new three-phase strategy and a full-time and part-time Response Force? Breen's books *Struggling for Self-Reliance* and *Disappointing the Dragon* and Langford's and Armstrong's theses contain the evidence to make a case for countering the CCP grey zone threat at home and overseas with a sovereign Australian strategy. Each book and thesis analyses the relevant literature and draws on primary and secondary sources to make arguments. The authors build the case for a complementary sovereign Australian strategy based on their practical operational experience, deep research and PhD-level exposition.

PhD-level authorship and a solid empirical foundation do not mean the book is a joined-up 'mega-thesis'. There are no footnotes or long, meticulous lines of academic argument to satisfy examiners. It is written for everyone interested in keeping Australia safe in troubled times. The authors invite readers to Google along the way to give confidence that they are not 'making stuff up'. The book includes a bibliography with links to assist admirers, sceptics, and detractors.

All for nothing?

Australia won't need a Let's Trade, not Argue strategy if the CCP calls off its grey zone campaign, returns to trading, and does not undermine the US-Australian alliance. The G20 meeting in November 2024 marked the peak of the CCP Phase 1 light grey campaign and the culmination of an uncharacteristic CCP charm offensive. Xi Jinping and Prime Minister Albanese declared that the China-Australia relationship had returned to levels not observed since the 2014 free trade agreement and Xi's warm speech to both houses of parliament.

The Australian government hopes that Xi will not insist on Australia's distancing itself from the US alliance by opposing President Trump's trade policies and other measures to constrain CCP illegal actions. Hope is not a strategy. Relying on diplomacy and negotiations to persuade the CCP to 'back off' in the grey zone is appeasement. Reliance on American and British promises for the 2030s and 40s ignores history and the realities of what will happen in the 2020s that now echo the 1930s. A complementary diplomatic deterrence strategy backed by a Powerful Owl Response Force is urgently required.

Contents

Abbreviations and Glossary

ABF Australian Border Force

AFP Australian Federal Police

AIF Australian Imperial Force

ALP Australian Labor Party

ANU Australian National University

ANZUS Treaty The Australia, New Zealand, United States Security Treaty is a 1951 collective security agreement initially formed as a trilateral agreement between Australia, New Zealand, and the United States. While the treaty has lapsed between the United States and New Zealand, it remains separately in force between those states and Australia.

ASD Australian Signals Directorate

ASIO Australian Secret Intelligence organisation

ASIS Australian Secret Intelligence Service

ASPI Australian Strategic Policy Institute

Attorney General's Attorney General's Department (Australian Government)

AUKUS Trilateral security partnership between Australia, the United Kingdom, and the United States was signed in September 2021.

BRI (PRC) Belt and Road Initiative: Partnerships with China in significant infrastructure projects, especially dual-military use like ports and airports. It promotes economic connectivity and cooperation between countries by expanding and improving transportation networks like roads, railways, ports, and airports. The stated aim is to enhance regional integration, increase trade, and stimulate economic growth. The unstated objective is to expand Chinese influence.

CCO Covert and Clandestine Operations

CCP Chinese Communist Party

CDF Chief of the Defence Force

CIA Central Intelligence Agency (US Government)

CIB Commonwealth Investigation Branch

CIS Commonwealth Investigation Service

clandestine conducted secretly to hide the fact that the activity is taking or has taken place

CMF Citizen Military Forces

CNA Computer Network Attack

CNE Computer Network Exploitation

CNO Computer Network Operations

CNS Chief of National Security

Country craft is a fleet of specially constructed vessels and modified trawlers called Snake Boats, identical to junks operating in the 1940s in the South China Sea (circa Pacific War).

CCP Communist Party of China

Covert extraordinary operations that hide the sponsor's identity or plausibly deny sponsorship

CSBA Center for Strategic and Budgetary Assessments (US think tank)

CSS Commonwealth Security Service (Australian circa 1930s)

Cyber actions are attacks by hackers to disrupt ICT systems that drive economies, essential services (water, power, sanitation, communications, public and private transport) and infrastructure.

DFAT Department of Foreign Affairs and Trade

Defence Department of Defence

De-escalation cycle Detection, 'staring' (continuous overwatch), analysis, decision, communication/ warning, and action, if required (also theory of application)

DHA Department of Home Affairs (Australian)

DHS Department of Homeland Security (US)

Direct effect Operations are the precise execution of activities anywhere and at any time. They employ small teams in hostile, denied, or politically sensitive environments to influence and de-escalate threats in the first instance and seize, destroy, capture, exploit, recover, or damage designated targets if warnings are ignored and escalation continues.

DSB Defence Signals Bureau

DSR2023 Defence Strategic Review 2023 (Australian Government)

DSU2020 Defence Strategic Update 2020 (Australian Government)

Electronic actions deny access to the Electro-Magnetic Spectrum [EMS] for communications – old-fashioned jamming – and threats to the reliability of devices that rely on the Global Positioning System (GPS).

EMS Electro-Magnetic Spectrum. Electromagnetic radiation according to frequency or wavelength. Although all electromagnetic waves travel at the speed of light in a vacuum, they do so at a wide range of frequencies, wavelengths, and photon energies. The electromagnetic spectrum comprises the span of all electromagnetic radiation. The electromagnetic spectrum, from the lowest to the highest frequency (longest to shortest wavelength), includes all radio waves commercial radio and television, microwaves, radar), infrared radiation, visible light, ultraviolet radiation, X-rays, and gamma rays. Nearly all frequencies and wavelengths of electromagnetic radiation can be used for spectroscopy.

EMSO Electro-Magnetic Spectrum Operations

Espionage CCP espionage steals defence, political, industrial, foreign relations, commercial, public sector or other information for China's advantage.

FBI Federal Bureau of Investigation (US Government)

FFP is a fictional anti-American and pro-China Forum for Peace and Prosperity in Australia

Foreign Affairs Department of Foreign Affairs and Trade (DFAT)

Foreign Interference ASIO task force that aims to discover, Threat Assessment track and disrupt foreign interference Centre in Australia

FSM Federated States of Micronesia

Ghosts are consummate professionals who represent the direct-action intelligence, informational, and precision-targeting dimensions of Response Force operations. They employ so-called 'ghost' technologies to conduct extraordinary operations in every domain.

Goblins are consummate professionals who will conduct extraordinary operations in the ICT, information, cyber and space domains.

GPS Global Positioning System

Grey zone is a contested political and economic battlespace between peace and war. Legal and illegal actions are used to achieve political objectives and gain control. *Activities designed to coerce countries in ways that seek to avoid military conflict and* calibrated to be under the threshold of what are considered acts of war

Guardians are consummate professionals prosecuting the indirect approach that encourages, guides, and empowers allies in Australia's near region to develop de-escalation strategies and response forces that suit their preferences for de-escalating in the grey zone.

Home Affairs Department of Home Affairs (Australian)

HUMINT Human Intelligence (information acquired by Australian spies and those Australia pays to spy)

ICT Information Communications Technology

IDTF Inter-Departmental Task Force

IMINT Image Intelligence (information gathered from aerial photography and satellites)

indirect enabling are activities that involve a combination operations of lethal and non-lethal actions taken by specially trained and educated teams that have a deep understanding of cultures and foreign languages, proficiency in direct effect de-escalation tactics, and the ability to operate with and alongside allied security agencies and other third parties in permissive, uncertain, or hostile environments if warnings are ignored and escalation continues.

Influence operations use information, psychological methods and relationships to change minds.

Information actions are the manipulation and dissemination of stories of destiny – propaganda – to influence people's attitudes and actions globally, not least the Russian and Chinese populations. (Russia and China) Australian information actions include public and covert information, propaganda and disinformation to shape CCP emotions, attitudes, motives, thinking processes and behaviour.

information dominance is the ability to collect, manage, and exploit accurate and self-serving inaccurate details more rapidly than anyone else.

ISIS Islamic State of Iraq and Syria

Joint Logistics Support facilitates joint logistics across the Force (PRC) services

Mossad Israel's armed intelligence service

MI5 Britain's domestic intelligence service

MI6 Britain's international intelligence service

MSC Motorised Submersible Canoes, called Sleeping Beauties, carried limpet mines that could sink two 10,000-ton ships if correctly placed to deploy off harbours before launching MSC operated by specially trained commandos to sink ships (circa Pacific War).

NSS National Security Strategy

NATO North Atlantic Treaty Organisation NATO's 32 members (in order of accession) are: (original 12 signatories in alphabetical order) Belgium, Canada, Denmark, France, Iceland, Italy, Luxembourg, the Netherlands, Norway, Portugal, the United Kingdom and the United States, (post-1952) Greece, Türkiye, Germany, Spain, (post-Cold War) Czechia, Hungary, Poland, Bulgaria, Estonia, Latvia, Lithuania, Romania, Slovakia and Slovenia, Albania and Croatia, Montenegro, North Macedonia, (post-Ukraine invasion) Finland and Sweden

NDS2024 National Security Strategy 2024 (Australian Government)

NEOC National Emergency Operations Centre

NEMA National Emergency Management Agency

NIC National Intelligence Community

NSC National Security Committee of Cabinet

ONI Office of National Intelligence

Operation Hornbill was a proposed series of SRD operations in the Pacific War in the South China Sea area, targeting the ports of Singapore, Saigon, and Hong Kong and inserting groups of agents to gather intelligence, enable resistance movements and conduct 'small, annoying pin-prick raids' against airfields, ports and railway infrastructure.

PLA (PRC) People's Liberation Army (Services: PLA Air Force, PLA Army, PLA Navy and PLA Rocket Force. Sub-Service Forces: Strategic Support Force and Joint Logistics Support Force)

PM Prime Minister

PM&C Department of Prime Minister and Cabinet

PPP fictional Party for Peace and Prosperity, facilitated by the CCP to campaign for power in Australia based on the PPP ending hybrid warfare and negotiating a China-Australia peace and prosperity agreement

PRC People's Republic of China

QUAD The Quad is a diplomatic partnership between Australia, India, Japan, and the United States committed to supporting an inclusive and resilient Indo-Pacific that is open, stable, and prosperous.

RAMSI Regional Assistance Mission to Solomon Islands

RF Response Force

RFR Response Force Reserve

SASR Special Air Service Regiment (UK and Australia)

SIGINT Signals Intelligence – information gathered by listening to others' telecommunication and penetrating ICT systems

SIC Signals Intelligence Centre

Snake boats See Country craft

SOA Special Operations Australia (circa 1940s)

SOE Special Operations Executive (UK circa 1940s)

Sovereign strategy is a self-reliant strategy Australia can implement independently with its own resources to keep Australia safe.

SRD (Aus) Services Reconnaissance Directorate (Australian circa 1940s). The SRD evolved into a fourth fighting force, a formidable armed intelligence service with aircraft, small boats, submersibles, and sporadic access to allied submarines and service aircraft.

SSF (PRC) Strategic Support Force – centralises the command and control of space and information operations (including cyber (defence, offence, and reconnaissance), electronic (interfering with or disrupting electronic and communications equipment, with an emphasis on jamming and anti-jamming) and psychological warfare, including public opinion and legal warfare).

SSF Network Systems Department (PRC) links purely military elements such as electronic warfare to elements such as cyber and psychological warfare, which also can have political, economic, and intelligence dimensions

SSF Space Systems Department (PRC) space launch and support, space surveillance, space information support, space telemetry, tracking and control, and space warfare

United Front (PRC) The United Front Work Department

UN United Nations

UNSW University of New South Wales

US United States

USSR Union of Soviet Socialist Republics

WMD Weapons of Mass Destruction

Part One – The Threat

What's up? What might happen? She'll be right, mate?

Part One argues that there is an urgent national security threat in the grey zone (**Chapter 1**). Xi Jinping and hardline CCP nationalists are aggressively seeking to dominate China's Asia Pacific neighbourhood and change the international rules-based world order to suit China's national interests. They are rapidly escalating pressure in the grey zone – a contested political and economic battlespace between peace and war – to coerce Asia-Pacific neighbours to sign agreements for peace and prosperity that formalise tribute state status.

Chapter 2 predicts what might happen if Australia does not have a strategy and the ways and means to deter a CCP grey zone escalation in the 2020s.

Chapter 3 argues that while the Australian government has recognised the threat and taken some action, these responses are not comprehensive enough. The government's National Defence Strategy 2024 (NDS2024) is a step in the right direction, but more needs to be done. Australia must be able to discourage the CCP from actions and effects that escalate from persuasion to coercion, from legal to illegal. The message is that Australia wants a trade-based relationship but will respond decisively and lawfully to unlawful activities.

CHAPTER 1

What's Up?

Introduction

The Chinese Communist Party (CCP) wants Australia and its Pacific Islands neighbours to become tribute states. A tribute state allows a more powerful state to tell it what to do. The CCP's objective is to sign agreements for peace and prosperity with Australia and its neighbours on China's terms, cloaked in mutual economic self-interest. The CCP prefers to take over through negotiation and economic and developmental incentives, as well as coercion if incentives are unsuccessful. Some may argue, 'Australia is already an American tribute state, so why not get a better deal with China?' Think again. There is a difference.

What does becoming a Chinese tribute state look like? Let's describe two examples briefly. Tibet became a tribute state in 1959 after signing a 'one country, two systems' agreement called the Seventeen Point Agreement in 1951 that guaranteed independent Tibetan political, economic, cultural and legal systems. It took eight years to undermine Tibetan sovereignty. Tibetan history since 1959 testifies to what becoming an occupied tribute state means. Google 'Chinese atrocities in Tibet.'

The people of Hong Kong also know what the CCP's tribute status feels like. They and the British Government agreed to a 'one country, two systems' agreement called the Sino-British Declaration in 1997, guaranteeing the independence of Hong Kong's political, economic, cultural and legal systems. The escalating grey zone campaign began in the early 2000s with bullying of so-called 'separatists'. The aim was to extinguish Hong Kong's growing liberal democracy in the name of loyalty, unity and Chinese patriotism and defer to the CCP. By 2019, the metaphorical democratic frog in the pot finally realised the water had become too hot for survival. It was too late.

The eruption of civil protests in Hong Kong in 2019, the arrest and detention of 15 pro-democracy dissidents in April 2020, and more arrests of over 300 democracy activists in July after the passage of more restrictive security legislation testifies to escalating CCP grey zone tactics. After denying the right of assembly and jailing the last of the 'separatists', Hong Kong became a Chinese tribute state. In 2023, a sham democracy emerged with only 20 per cent of seats in District Councils elected directly from CCP-screened representatives. The local Chief Executive runs Hong Kong at the direction of Beijing. Google 'Chinese Communist crackdown in Hong Kong'.

Are Tibet and Hong Kong false comparisons with Australia because both countries were 'more or less' in the Chinese orbit anyway? Was the CCP just securing borders? Is the tension with Taiwan the same: a border assertion issue, not a grey zone bullying campaign leading to a takeover? Indeed, isn't it up to the Taiwanese to sort out how to deal with the CCP's ideological anxieties about its borders? Surely, the CCP militarisation of islands in the South China Sea and aggressive behaviour towards northern and southeast Asian neighbours are further examples of China securing its borders – defensive, not offensive? Why should a drop of Australian blood be spilt over a long-running Chinese civil war or China's disputes with its Asian neighbours?

Do Asian neighbourhood disputes have anything to do with Australia and its Pacific Islands neighbours located thousands of kilometres away from the Chinese homeland? Surely, the CCP will stop at the First Island Chain (the Kuril Islands, Japanese archipelago, Ryukyu Islands, Taiwan, northern Philippines, Borneo and the Malay Peninsula) for China's maritime border security. Where is the evidence of CCP ambitions to create tribute states further east and south, including Australia and the Pacific Islands? Surely, the CCP has little interest in imposing control on Australia, an important trading partner, and the Pacific Islands nations, minnows in the international community. Surely, the CCP only seeks their self-entitled sphere of influence in Asia. Think again. But wait a moment while we tell you about the grey zone before proving that there is 'danger on our doorstep' to borrow from the late Jim Molan's book title.

The Grey Zone

The term 'grey zone' applies to what nations do to each other legally, illegally and clandestinely between the 'white' of peace and the 'black' of war. Grey zone campaigns involve non-violent and violent actions by predatory nations that are initially diplomatically, politically and economically persuasive. Suppose the persuasive 'light grey' first phase (Phase 1) does not work. In that case, escalation involves coercive and often harmful actions while remaining below the threshold of what is usually considered acts of war. Imagine a nation or organisation as a frog in a pot. A grey zone campaign aims to heat the water around the frog slowly without it knowing or, if it feels the extra heat, dismiss the temperature rise as not harmful or even manageable. Eventually, the water becomes too hot for the frog to survive.

A grey zone campaign is a veiled war when there is seeming peace. It is a gradual and secret escalation of diverse activities to influence and persuade nations, organisations and individuals to cooperate in the interests of the country or subnational group prosecuting it. The initial activities (Phase 1) are meetings, visits and negotiations for free trade and other diplomatic agreements that mask espionage, political subversion, cyberattacks, propaganda and interference. Suppose initial non-violent persuasion and incentives don't work. In that case, activities become more coercive (Phase 2) and eventually escalate to hybrid warfare (Phase 3), which involves violent actions by deniable unmarked military forces, armed spies, recruited criminals, mercenaries and provocateurs – grey zone combatants.

Contemporary grey zone campaigns integrate political, economic, cyber, electronic (e.g. jamming communications) and information actions (propaganda, often in social media). Typically, campaigns oscillate back and forth through three electronic and subversion phases.

Phase 1. Political persuasion and economic incentives and pressure.
Phase 2. Political subversion, economic pressure and cyber/electronic attacks.
Phase 3. Political subversion, economic pressure, cyber/electronic attack and hybrid warfare.

The label 'hybrid' encompasses centuries-old ways of war on the sea, land, and in the air, but it is enhanced with cyber, artificial intelligence (AI), information, and drone technology. At the time of writing in 2024, the Iran-backed Houthi rebels are giving Western allies a masterclass in low-investment, high-return harassment of commercial and military shipping in the Red Sea to demonstrate how Iran's low-level hybrid warfare works. Iran's sponsorship of Hezbollah in southern Lebanon and Hamas in Gaza are examples of using proxies for Phase 3 hybrid warfare against Israel. (Google Iran's regional armed network) Israel's assassinations of Arab leaders and nuclear scientists, exploding pagers and hand-held radios and the bombing of facilities suspected of developing Weapons of Mass Destruction (WMD) are examples of special operations in the grey zone. (Google Hezbollah Pager Explosions)

Preparations are secretive, denied and disguised. Those prosecuting a hybrid war use their special forces and armed spies to fund and recruit locals, typically criminals, malcontents and misguided activists and influencers (useful idiots), to protest, intimidate, and eventually take up arms against their society as 'patriots' fighting for change. By this time, a specially recruited political group advocates for change, usually closer relations with the nation sponsoring hybrid operations and works hard to remove the current government. That group is positioned to take power when the existing government, institutions, and civil society cannot function effectively after years of subversion and destabilisation.

The Russians have provided masterclasses of hybrid warfare in their neighbourhood with short, sharp and nasty campaigns in Estonia and Georgia in 2007/8 and a successful takeover of Crimea in 2014 after a brief winter campaign (more about these campaigns in the next chapter). The campaign in the Donbas region of Ukraine intensified soon after Crimea succumbed. The United States and NATO sat on their hands. European nations that were dependent on Russian oil and gas did not want to pick a fight during winter. The current conventional war between Russia and Ukraine testifies to how a hybrid war can escalate when its sponsoring state becomes frustrated. This conflict has become a proxy war between the US and Russia, reminding us of similar conflicts in Africa during the first Cold War. The deployment of North Koreans to assist the Russians adds a new Cold War dimension to the tragedy.

Attitude to China

Before proving that Australia and its neighbourhood are CCP takeover targets, we need to briefly clarify our attitude toward China to remove the temptation for critics to 'cancel' us as racists and irrational anti-Communists. Our message is, 'Let's trade with everyone for mutual benefit and argue with no one.' However, if the CCP bullies Australia, there must be uncomfortable consequences so relations can return to trade and not argument.

Australia has a significant trading relationship with China. Supply chains to and from China create co-dependencies between Australia's economy based on China's willingness to supply manufactured goods and China's economy based on Australia's willingness to export raw materials, especially iron ore, lithium and liquified natural gas. Using the marriage analogy, Australia should be 'married' to China for trade, not for love. Australia and the CCP will never love and admire each other politically because both are different and always will be. Those political differences should not stand in the way of mutually beneficial trade, the relationship's bedrock. Australia should mind its own business and not comment on CCP behaviour unless it is illegal, harmful, or interferes with trade. There is no place for rude, nasty or argumentative behaviour between trading partners.

Mutual respect is essential. Transactions between the Australian and Chinese peoples are warm, respectful and mutually beneficial. Australia educates thousands of Chinese students in Australia's university sector and welcomes thousands of Chinese tourists annually. Over 1.2 million Australians of Chinese heritage live in Australia. Their cultural practices may be Chinese, while they enjoy Australian lifestyles and join other Australians in 'Aussie' national pride. Thousands of Australians visit and admire Chinese culture, landscape, history, architecture, and achievements annually.

There is an immense reservoir of goodwill. Australia should make every effort to continue to foster good relations, especially in the face of self-serving provocations from CCP hardline nationalists. We do not equate the Chinese people and their corporate and civil society leaders with the aggressive behaviour of their current political rulers. The book's recommended de-escalation strategy encourages Xi Jinping and his hardline CCP nationalists to think twice about their grey zone campaign in Australia and its neighbourhood. It enables moderate CCP, corporate leaders, and ordinary Chinese citizens to make a case to return to mutually agreeable trade relations and respectful acknowledgement of political differences.

The authors have written this book in response to what a reasonable person would agree is threatening, sometimes described in commentary as 'bullying' behaviour and the assertive agenda of CCP nationalist hardliners in Australia and its near region. If the situation were reversed, CCP hardliners would not tolerate the Australian Government prosecuting a grey zone campaign in China or against Chinese national interests elsewhere. Though it is naïve to evoke this 'Golden Rule' for a severe discussion about international relations and 'pseudo-war', our messages are, 'Do unto others what you would want others to do unto you.' 'Leave us alone, and we will leave you alone.' 'Let's trade, not argue.'

We recommend a firm, lawful and fair response to the CCP's illegal activities and intimidation that accords with the sentiments and values of Australian and Chinese citizens and their leaders. Still, we advocate increasing uncomfortable consequences for subversive and threatening illegal behaviour. Our strategy anticipates the temptation for the CCP to escalate if Australia does not comply with encouragement to distance Australia from the United States under its 'America First' President. Eventually, the CCP will ask Australia to sign agreements for peace and prosperity on China's terms.

We believe the basis of the relationship with China, as with all nations, should be mutually beneficial trade – a marriage based on trade, not an abusive relationship based on plotting and subverting to gain more control. So, what are we concerned about?

The Big Picture
China and Russia have started a new grey zone Cold War, emulating

the political war of the first Cold War. Let's not debate the reasons or play the blame game. It just 'is'. During the first Cold War, a senior American diplomat, George F. Kennan, defined political warfare as 'the employment of all the means at a nation's command, short of war, to achieve its national objectives', essentially a grand strategy. Russia's unopposed annexation of Crimea in 2014 and China's uncontested building of island military bases in the South China Sea are measures of their early success. Ross Babbage, an eminent Australian strategist, opines:

> The political warfare undertaken by the regimes in Russia and China is, at its core, driven by the obsessions of Vladimir Putin and Xi Jinping to protect their personal rule. These leaders feel deeply threatened by the liberties and practices prized by liberal democracies.
>
> Hence, in order to defend themselves, rally domestic support, keep their enemies off-balance, and weaken and potentially overthrow democratic states, *they have refined powerful versions of political warfare as a means of progressing their interests at relatively low cost and risk* [authors' emphasis]
>
> They appreciate that by operating aggressively and in a nimble fashion in the grey zone between the Western conceptions of peace and war, they are exploiting a substantial advantage over the United States and its allies, who are more traditionally minded, conventionally structured [navy, army and air force], and bureaucratically sluggish.
>
> ... Moscow and Beijing employ a much wider range of instruments, many of which involve highly intrusive intelligence operations and deeply subversive espionage, cyber, military, and other active measures to disorientate, distract, confuse, coerce, undermine, and potentially cause the collapse of targeted societies.

Clive Hamilton, an eminent Australian public intellectual, has put together convincing evidence exposing the CCP's international campaign:

> The Chinese Communist Party is determined to transform the international order to shape the world in its own image *without a*

shot being fired. [authors' emphasis] Rather than challenging from the outside, it has been eroding resistance to it from within by winning supporters, silencing critics and subverting institutions. ... and while many in the West remain reluctant to acknowledge this, democracies *urgently need to become more resilient if they are to survive.* [authors' emphasis] ... As Beijing is emboldened by the feebleness of resistance, its tactics of coercion and intimidation are being used against an increasingly broad spectrum of people.

Australia is a member of a Western alliance of nations that seeks to preserve the rules-based global order, uphold the United Nations Charter, and support liberal democracy and human rights. This membership and a continuing closeness to the United States obligates Australia to comprehend China and Russia's ambitions. Vladimir Putin and Xi Jinping know that Australia bats for the US team that opposes Russia and China's self-entitled aspirations to dominate their desired spheres of influence.

How does Australia bat for the US team? Traditionally, when the United States goes to war, Australia does so as well, as in Korea, Vietnam, Iraq, and Afghanistan. Still, Australia only does so with modest contributions – a few ships, an infantry battalion or two, some special forces and a few aircraft. Those are niche contributions at best. Some like to call it 'Punching above our weight.' Others call it letting the Americans carry the weight of their strategic choices. Australia is another allied flag flying on international US battlefields, not a substantial military partner supporting a rules-based world order. Australia is an appreciated 'nice to have', not a 'must have' for US-led military interventions.

Since the end of the Pacific War in 1945, most of Australia's overseas military operations have been 'down payments' on an American strategic insurance policy. In effect, Australia wants the United States to repeat its defence of Australia during the Pacific War in return for these down payments. Australian governments assume the Americans will be interested in doing so again because American presidents have consistently affirmed the Asia-Pacific region's importance and the US-Australian alliance's value.

But will the Americans bat for us in the same way they did in 1942 after the Japanese attack on Pearl Harbour in December 1941? Japan's attack on the US Navy, followed by its southern thrust into Southeast Asia and the Darwin bombing campaign, quickly made Japan a mutual enemy. Though the priority was defeating Hitler in Europe, the United States deployed significant resources to Australia for a combined allied campaign to defeat Japan.

Initially, the Americans appreciated the will and weight of Australia's effort in New Guinea to stop Japanese land forces. Fortunately, the Americans had enough left of their navy after the Pearl Harbour attack to defeat the Japanese at Midway and in the Coral Sea in 1942. The Japanese were on the back foot after that. Notably, General Douglas MacArthur, the American Commander-in-chief of allied forces, left his Australian allies behind to mop up isolated Japanese forces during the last two years of the Pacific War while he and his American forces 'island hopped' towards Japan.

A re-elected President Trump creates uncertainty about how America might bat for Australia. At his inauguration in 2025, he affirmed an America First approach to foreign policy and trade. During his first term, he assessed that other NATO members were not pulling their weight, calling on each of them to 'finally contribute their fair share'. His actions in 2017 led to commentary that he was 'throwing Australia and its allies under the bus' to achieve his America First foreign policy. President Trump likes private one-on-one meetings with dictators like Xi Jinping, Kim Jong Un, and Vladimir Putin.

In 2018, President Trump negotiated a 'ceasefire' with Xi Jinping after a brief trade war with China. Despite it not holding in 2019, this deal and the Abraham Accords between Israel and the United Arab Emirates, Bahrain, and Morocco testify to Donald Trump's lifelong habit of deal-making, evident in his commercial life and dramatised in the first 14 seasons of the NBC television series The Apprentice. Donald Trump has threatened 60 per cent tariffs on Chinese exports to the United States. Excusing the pun, under a new Trump administration, by 2026, US trade interests with China may trump American strategic commitments to protect its Asia-Pacific allies from CCP grey zone pressures.

Fast forward to the late 2020s and mid-2030s, the probable span of Xi Jinping's life-long reign, noting that Mao Zedong died at the helm at 90. Xi Jinping turned 71 in June 2024. Will he wait much longer to change the world? A Chinese electronic Pearl Harbour combined with a long-range missile surprise on US bases in Asia would galvanise another combined alliance effort. But Xi and the CCP understand history as well as we do. Why would the CCP expose Chinese armed forces to American firepower if they can achieve the same aims without firing a shot? Furthermore, why would China and the United States risk a conventional war that might escalate a nuclear exchange? At worst, they might fight proxy wars through allies, as Russia and the United States have done in Ukraine.

The US Navy, Army, and Air Force will not deploy to counter the CCP grey zone campaign against Australia. The ADF is not suited for this mission either because China will only deploy its navy, army, or air force to intimidate, not to invade. However, the CCP may 'roll the dice' as the Japanese did and surprise the United States and its allies with a sudden attack. In his 2023 book *Danger at Our Doorstep*, Jim Molan describes this electronic and missile Pearl Harbour to get his readers' attention before recommending massive spending on Defence. This book does not argue against being prepared for this possibility. Still, it strongly advocates countering what is happening now and preparing for what might happen next in a hi-tech malicious grey zone Cold War.

The CCP continues to build Chinese armed forces, especially its navy and missile capabilities, to eventually 'overmatch' the US Navy in Asia-Pacific. Time will tell if the CCP is up for territorial conquest. However, escalation in the grey zone will happen to test President Trump's resolve. It is the preferred pathway to dominance because it bypasses navies, armies, and air forces to subvert political, economic and social systems in homelands. Trump may tell allies to resist in the grey zone as their sovereign responsibility, not an American one.

The question is whether the United States will go to war to protect Australia and the Pacific Islands unless American national interests are at stake. Donald Trump promised to avoid a Third World War during his presidential campaign. Taiwan may be the 'canary in the coal mine' or possibly the Philippines before Taiwan, for determining US interests in defending Asia-Pacific allies. Tension between the United States and China over Taiwan's sovereignty may escalate or subside during Donald Trump's second term. He may 'do a deal', leaving Taiwan alone in the grey zone and other US allies wondering where they stand if the CCP escalates against them.

President Trump is unlikely to 'put boots on the ground' or put US naval vessels in harm's way to save Taiwan from invasion. Alternatively, Taiwan could become Asia's Ukraine, receiving massive US armaments, ships, submarines, and aircraft to fight a 'forever' war with China. Still, Xi Jinping is unlikely to invade anyway. He will most likely 'turn up the heat' in the grey zone after he assesses Trump's commitment to war to save Taiwan.

Takeover target?
The inconvenient truth is that despite Australia's lucrative trading relationship with China and the warmth of the relations and mutual admiration between the Australian and Chinese peoples, the CCP wants more control. Where's the evidence? We thank fellow Australians David Kilcullen, Clive Hamilton, Alan Dupont, Peter Jennings, Peter Hartcher, Peter Connolly, John Fitzgerald, Alex Joske, Ross Babbage and others, as well as New Zealander Anne-Marie Brady for 'blowing the whistle' on the CCP's grey zone campaign in Australia and its regional neighbourhood (see the Bibliography).

Clive Hamilton is explicit and specific about the CCP threat in Australia in his book *Silent Invasion*. Peter Hartcher's *Red Flag* does the same. Peter Connolly's PhD thesis, *Statecraft and Pushback*, reveals the CCP's grand strategy in Melanesia right on Australia's doorstep. No one has challenged these authors credibly. Their comprehensive analyses identify the grey zone frontline – the political system, the economy, trade, corporations, the education system, the media, entertainment and cultural organisations and civil society more generally. In sum, these authors whose books and articles are listed in the Bibliography provide compelling proof that something sinister is going on and, in our opinion, will escalate. Our message is,' Let's trade, not argue.' The CCP message appears, 'Let's trade, but on our terms with a tribute state.'

There continues to be a lively, sometimes sensationalised, national discussion in the Australian media about CCP surveillance, infiltration and influence in Australia. The Dastyari affair dominated the discourse in 2016 and 2017. Late in 2019, there was a flurry of interest in a Chinese defector seeking asylum in Australia. His interviews on Channel 9's *60 Minutes* TV program confirmed CCP's interest in increasing its influence in Australia. But there are doubts about the defector's authenticity. In May 2024, the ABC *Four Corners* program interviewed a former CCP secret police agent who blew the whistle on CCP intimidation of dissidents in Australia and elsewhere.

The Murdoch press questioned the former Victorian Premier, Daniel Andrews, and his government for signing agreements related to the Chinese Belt and Road Initiative. Articles point to CCP agents of influence and imply that former politicians promote CCP interests in Australia. Greg Sheridan, *The Australian* newspaper's foreign editor and prominent media commentator, has written numerous articles warning of CCP influence in Australia and the surrounding region. On 27 April 2024, *The Australian* newspaper's front-page headline was, 'China's invisible invasion', referencing US reports of CCP infiltration of ICT systems worldwide, including Australia.

Late in June 2020, the media reported on ASIO officials and AFP officers raiding the home and NSW parliamentary office of ALP MP Shaoquett Moselmane and the home and office of John Zhang, one of Moselmane's staff, on suspicion of their complicity in CCP influence operations. Significantly, these raids were the first conducted under provisions of new legislation aimed at quelling foreign interference. Clive Hamilton commented that there was further evidence of CCP infiltration of the NSW ALP. *The Australian* newspaper headlined that ASIO was investigating the CCP infiltration of the NSW Parliament. This contested evidence suggests a lively interest in the media of any signs of illegal CCP influence. Still, conclusive evidence in the media is elusive.

What does Australia's National Intelligence Community (NIC) say? Indeed, it's up to them, not academics and journalists, to detect and warn about what threatens Australia. All Australians should be grateful to the Australian Security and Intelligence Organisation

(ASIO), the Australian Signals Directorate (ASD), and the Australian Strategic Policy Institute (ASPI) think tank, possibly informed by the NIC, for further revelations. Please consult the Bibliography to learn more about the NIC's warnings. Mike Burgess, Director General, ASIO, was precise in 2020:

> The level of threat we face from foreign espionage and interference activities is currently unprecedented. *It is higher now than it was at the height of the Cold War.* [author's emphasis]… There are more foreign intelligence officers and their proxies operating in Australia now than at the height of the Cold War, and many of them have the requisite level of capability, intent and persistence to cause significant harm to our national security. But the character and focus of that espionage activity will continue to evolve.

Though Burgess, for political reasons, cannot call out the CCP, let's be clear: the primary organisation conducting unprecedented levels of espionage, cyber disruption, political interference, influence operations and intimidation of citizens in Australia is the CCP, with India looming in second place. Burgess doubled down on his 2020 warning in his 2021, 2022 and 2023 annual assessments, pointing to a steady and alarming escalation in espionage and interference, including a plot to influence the 2022 Federal election illegally. In 2024, he called out an unnamed ex-politician for 'selling out' Australia and specified that an unnamed nation's 'A Team' intelligence service was targeting Australia. He meant China.

In 2023, Burgess risked his job by warning the Albanese government not to trade Australia's security for economic and political gain. He restated that Australia faces 'an unprecedented challenge from espionage and foreign interference' damaging the nation's 'security, democracy, sovereignty, economy and social fabric'. ASIO is on the frontline against 'sophisticated foreign adversaries effectively unconstrained by resources, ethics and laws.' We wish the men and women of ASIO well and thank them for their service, but the government needs to back them up with capabilities to de-escalate the threat they are warning Australians about.

The Australian Signals Directorate (ASD) issues annual warnings about 'state actors' conducting cyber-attacks against Australia. Let's be clear again: it's primarily the CCP orchestrating cyber-attacks as part of its grey zone campaign. Aside from organised criminals and mischievous Russians and others practising their skills, who else is interested in conducting cyberattacks against Australian institutions, government departments and businesses?

Don't be tempted to think that all this cyber-attack stuff will not have consequences for ordinary Australians, their families, and their communities. It is an invisible invasion of Australian defences and preparation for 'darker grey' escalation. The ASD warned on 20 March 2024 that the Chinese state-sponsored hacking group Volt Typhoon targets Australia's infrastructure and ICT systems. Mike Burgess warned in 2024 of 'multiple attempts to scan critical infrastructure'. *The Australian* characterised these efforts as 'the 'electronic equivalent' of Chinese commando groups putting bombs underneath bridges or on high-voltage pylons to blow them up during the war.' The Chief Executive of the Cyber Security Cooperative Research Centre, Rachael Falk, says Chinese hackers 'sit in wait, ready to attack in the event of a major conflict.' In other words, the CCP is infiltrating Australia's ICT systems to disrupt them and Australia's way of life to force political concessions if the Australian Government cannot be persuaded to sign an agreement for peace and prosperity on their terms.

'The intention is not to steal information [espionage] but to control critical systems' [communications, energy, transport, healthcare and water], according to *The Australian* on 27 April 2024. Imagine the distress if smartphones and other Internet-connected devices cease to function properly. If online banking goes down. If online government services break. If the IT systems coordinating food supplies to Woolies, Coles and Aldi go down. The 2020 stampedes for toilet paper during the COVID pandemic will be nothing compared to the behaviour of Australians if food and petrol supply chains malfunction and electricity and water supplies are disrupted. The 2017 Independent Intelligence Review confirmed:

... More generally, the cyber domain will likely feature even more prominently than it currently does in attempts to undermine economies, societies and national governments. It offers a relatively inexpensive but potentially effective way of achieving a wide range of effects – from influencing political processes to disrupting financial systems and key aspects of national infrastructure.

Though no grey zone strategy is identified, the Department of Defence (hereafter Defence) is aware of the grey zone. Australia's *2020 Strategic Update* states:

> These [grey zone] tactics are not new. But they are now being used in our immediate region against shared interests in security and stability. They are facilitated by technological developments, including cyber warfare.

The NDS2024 warns of 'grey zone activities', 'the prospect of coercion', and increasing foreign interference, espionage, and cyber-attacks. However, the five tasks given to the ADF do not include countering grey zone activities. Instead, the focus is on the ADF becoming 'more capable of the impactful projection of military power.'

The Australian Secret Intelligence Service (ASIS) Director-General does not issue annual threat assessments. Still, ASIS is engaged in the grey zone, attempting to detect and disrupt threats to Australia's national interests 'upstream' overseas, while ASIO does so 'downstream' in Australia. Unlike ASIO, which can inform law enforcement agencies to act against law-breaking and ADF special forces to thwart violent terrorist acts, ASIS does not have the authority to employ armed operatives to counter grey zone threats to Australia's national interests overseas. The CCP has the escalation initiative there – metaphorically, it can 'punch' freely in the international grey zone. At the same time, Australia is left to complain to the World Trade Organisation and the United Nations while it affirms the US alliance as the bedrock of Australia's national security.

The United Front Work Department
Political and social infiltration is the work of the CCP United Front

Work Department, a network of party and state agencies responsible for influencing groups outside the party. The CCP's role in this system's activities, known as 'united front work', is often covert and deceptive. ASPI's Alex Joske summarises nicely:

> The united front system's reach beyond the borders of the People's Republic of China (PRC) – such as into foreign political parties, diaspora communities [such as Australians of Chinese heritage] and multinational corporations – is an exportation of the CCP's political system. This undermines social cohesion, exacerbates racial tension, influences politics, harms media integrity, facilitates espionage, and increases unsupervised [illegal] technology transfer.

Mike Gallagher of the US Senate's Select Committee on the CCP is more explicit about the United Front system.

> United Front Work is the party's strategy to fight unrestricted political wars through three main tactics: silencing criticism of the regime globally, promoting propaganda abroad, and manipulating foreign institutions through clandestine and outright illegal operations. Perhaps the best summation of the united front strategy I've heard is simply three words, 'making idiots useful,' meaning co-opting any individual or organization to advance the goals of the Party.

Here are some Australian examples and warnings. In 2016, there were disclosures of impropriety between Australian Labor Party [ALP] Senator Sam Dastyari and Chinese businessman Huang Xiangmo. On 9 June 2020, ASPI released Alex Joske's analysis, *The Party Speaks for You*, containing compelling evidence of CCP political warfare in Australia. He describes the 'united front system', a grouping of agencies, social organisations, businesses, universities, research institutes and individuals carrying out united front work to persuade the Australian political system, the corporate sector and civil society to favour CCP interests. In 2022, ASPI published a detailed description of CCP United Work Front infiltration into every Australian State and territory (see the Bibliography).

The United Front can be nasty. Peter Hartcher in *Red Flag, Waking up to China's Challenge*, warns of China's 'insatiable appetite' for influence in Australia. He highlights examples of CCP agents harassing prominent citizens and their families that echo the intimidation of Anne-Marie Brady and her family in New Zealand (Google Anne-Marie Brady intimidation). He reports on CCP harassment of journalist John Garnaut and his wife after Garnaut authored a classified report on CCP interference in Australia for the Turnbull Government in 2015. It illustrates that the CCP likes to 'get up close and personal'. Garnaut's report prompted the Morrison Government to introduce legislation countering CCP covert intrusions in 2018.

Another example is the continuing campaign against Australian Olympic swimmer Mack Horton and his family, which was exposed in the media in 2020 (Google Mack Horton intimidation). Horton 'called out' Chinese swimmer Sun Yang, a three-time Olympic Gold medallist and 11-time world champion at the 2016 Olympics, for being a drug cheat and refused to stand on the dais with him at the 2019 World Swimming Championships. Interestingly, Brazilian authorities were sufficiently concerned about the safety of Horton and his parents in Rio de Janeiro in 2016 to assign armed special forces commandos to keep them safe. Back in Australia, the Horton family found broken glass in their swimming pool, their cars sabotaged, and Mack's father's business cyberattacked. The Horton family reportedly received regular briefings from ASIO officials for their safety in 2020, and these briefings have continued to the time of writing in 2024, even after Sun Yang's suspension from international swimming competitions for eight years for proven drug testing offences.

Another example of CCP intimidation in 2020 was the physical attack on and continuing harassment of University of Queensland activist student Drew Pavlou for his support for the pro-democracy movement in Hong Kong (Google Drew Pavlou intimidation). Pavlou has since called out the University of Queensland for hosting a CCP-funded Confucius Institute and permitting CCP agents to mobilise Chinese students to shut down peaceful pro-Hong Kong democracy protests on campus violently.

John Garnaut, Mack Horton and Drew Pavlou are publicised

examples of the CCP intimidating Australians in their own country. It is worth pondering how many citizens CCP agents threaten directly or through criminals and radicalised Chinese student proxies and if numbers are growing (Google Chinese interference in Australia). More particularly, how might Australia detect, deter and de-escalate these illegal activities that must be dreadfully distressing for victims and their families? The book's proposed de-escalation strategy would draw a line on this type of intimidation and create uncomfortable consequences for perpetrators – grey zone combatants.

What about Australia's regional neighbourhood? Is the CCP trying to distance Australia from being the Pacific Islands' most capable and supportive neighbour? Peter Connolly's thesis, Anne-Marie Brady's books and articles, and think tank analyses prove that the CCP targets Australia's regional neighbourhood (Google Chinese influence in the Pacific Islands). China signed multiple bilateral agreements with almost a dozen Pacific Island nations during its foreign minister Wang Yi's visit to the region in 2022 (Google Wang Yi visits the Pacific Islands). Most related to the economy, health, disaster response, and technology. He failed to sign up the countries he visited to bilateral security cooperation agreements related to policing, cybersecurity and ocean mapping. He had succeeded in Solomon Islands earlier (Google China-Solomon Islands security agreement). This agreement was troubling because it enabled Chinese police advisers to train Solomon Islands police, a controversial law enforcement agency implicated in political manipulation and corruption in the 2000s before the Regional Assistance Mission to Solomon Islands (RAMSI) intervention in 2003. Commenting on Wang Yi's efforts to persuade Pacific Island leaders to sign security pacts, Michael Shoebridge, ex-director of ASPI's Defence, Strategy and National Security Program, observes:

> The big thing to take away from Beijing putting such a regionwide security pact together is that this is now an overt statement of China's ambition to play a direct security role within small Pacific states and across the South Pacific.

David Panuelo, the president of the Federated States of Micronesia

(FSM), wrote a letter to FSM leaders providing extraordinary details on Beijing's political warfare and grey zone activity in the country in early 2023 (Google David Panuelo letter). He is a lone voice among Pacific Islands leaders who see the benefit of playing China off against Australia to secure more aid for their nations. In November 2023, the Special Envoy for Pacific Island Countries Affairs of the Chinese Government, Qian Bo, congratulated the Cook Islands on hosting the third Belt and Road Forum for International Cooperation after attending the Pacific Islands Forum meeting in Avarua, the Cook Islands capital. There were efforts to exclude any mention of Taiwan in the 2024 Pacific Island Forum communique. The CCP Belt and Road Initiative (BRI) is not just an economic strategy, but a potential threat to the region's sovereignty. The BRI is about partnerships in significant infrastructure projects, especially dual-military use like ports and airports, which could shift the balance of power in the Pacific.

China has just opened its fifth major Antarctic research facility to the south of Australia (Googe China and Antarctica). The Qinling Station is on Ross Island and will be staffed year-round. Ostensibly intended to support research, the facility may have 'dual use' capabilities that could support or threaten Chinese force projection operations. This threat undermines Australia's geographic strategic depth and important maritime lines of communication. Is this a coincidence or a larger strategy to quietly surround Australia?

Let's wrap up 'What's up?' before discussing 'What might happen next?' The China threat debate in Australia varies. What's up? Is it real? Is it dangerous? Is it unstoppable? Corporations, groups, and individuals sympathetic to CCP interests discredit those warnings of tribute-state intentions. Some corporate leaders call for everyone to 'settle down' and continue a lucrative trading relationship. Reputable experts – doves – call for a new conciliatory and collaborative narrative for the China-Australia relationship. There is plenty of evidence in 2024 that this approach was working. Xi Jinping and Anthony Albanese's public statements at the November G20 meeting are examples. Albanese's emphasis that one in four jobs in Australia relies on trade is timely. Other experts – hawks – advise Australia to acquire more submarines, missiles, and strike fighters to oppose a Chinese invasion fleet in the

Indonesian-Melanesian archipelago (Google Hugh White on China invasion). No one is arguing for a Let's Trade, not Argue de-escalation strategy with Powerful Owl capabilities to counter the CCP in the grey zone to reduce the CCP's temptation to bully and seek more control.

The CCP's political war with the world is relevant to Australia. Firstly, it would be stupid to misread or underestimate China's political encroachments in Australia's near region and its silent and invisible invasion of the Australian homeland. The penetrative political operations against democracy in Hong Kong and Taiwan are instructive. Secondly, Australia should not depend on allies to deter CCP intentions. Thirdly, it is Australia's national responsibility to deter political warfare while at the same time maintaining trade relations – a firm but fair and ethical de-escalation strategy.

No one will ever know beyond a reasonable doubt what Xi Jinping and the CCP's intentions are for Australia and the Pacific Islands because the essence of the CCP *modus operandi* is to disguise motives, ways, and means. Hindsight will be the 'exact science' for discovering the truth. The truth beyond a reasonable doubt is that history is replete with nations, peoples and alliances that failed to heed strategic warnings about totalitarianism. Consequently, they were surprised and defeated.

Let's discuss what might happen if Australia has no grey zone strategy.

CHAPTER 2

What might happen?

Introduction
This chapter discusses what China and Russia have been doing and what might happen in Australia based on what has happened elsewhere. The next chapter will argue that Australian responses to the CCP grey zone campaign, such as new laws against espionage and foreign interference, are tactical and responsive but do not anticipate what might come next. An Australian 'She'll be right, mate' attitude is dangerous and complacent.

The Russians gave a grey zone masterclass in Crimea in 2014 and later in the Donbas region before escalating to invading Ukraine in 2021. The cooperation between Russia and China means that China can and may do the same in its neighbourhood when the CCP deems the time is right. Readers need to know what this escalation may look like and how it will impact them, their families, and their communities to justify the book's Let's Trade, not Argue strategy backed by a Response Force.

The Big Picture
While the United States and its allies focussed on destroying terrorist networks and suppressing insurgencies in the Middle East and Afghanistan after 9/11, China and Russia took to the grey zone. Echoing the rhetoric of the first Cold War, Xi Jinping and Vladamir Putin spoke of China and Russia's entitlements to spheres of influence based on the historical empires of the USSR and the Middle Kingdom. They accused the United States of stifling Chinese and Russian destinies after China's two centuries of humiliation and the breakup of the Soviet empire after the end of the Cold War, followed by the expansion of NATO. They began a political and information war that, unlike the 1950s and 1960s, was powered by the Internet, data exploitation and manipulation, cyber-

attacks and a susceptible global media.

China and Russia have developed state-of-the-art information, cyber, and electronic capabilities. What are they? Information actions manipulate and disseminate stories of destiny – propaganda – to influence people's attitudes and actions globally, not least the Russian and Chinese populations. Cyber actions attack ICT systems that drive economies, essential services (water, power, sanitation, communications, public and private transport) and infrastructure. Electronic actions deny access to the Electro-Magnetic Spectrum [EMS] for communications – old-fashioned jamming – and threats to the reliability of devices that rely on the Global Positioning System (GPS).

Russian and Chinese grey zone campaigns target populations, not military forces. Cyber and electronic attacks target economic activity and everyday life to coerce people into paying attention and then acting in Russian and Chinese interests under the guise of mutual peace and prosperity, even patriotism. Russia and China sub-contract and deny grey zone operations. They characterise any discovered actions as responding empathetically to the calls from oppressed patriotic Russians and grievance groups living in Ukraine and the Baltic States and the oppressed people of Africa, the Middle East and Asia. China steadily creates military bases on contested atolls and islands in the South China Sea as the right of the Middle Kingdom to defend itself. Xi Jinping is restoring an entitled dominance after centuries of humiliation and European and American exploitation.

The Russians attack people's morale through their smartphones, a click and a few idle seconds, seeding misinformation and intruding with personal messages about individuals' circumstances, preferences and fears. They have weaponised social media and revived the Cold War use of proxies to do their dirty work. They send in special forces – so-called little green men – and mercenaries, like the Wagner Group, to mobilise hostile groups, assassinate individuals and concoct civil unrest among their neighbours, especially in Ukraine and further afield in the Middle East and Africa (Google Wagner Group). The number of Wagner Group mercenaries swelled from a few thousand to an estimated peak of 50,000 'dogs of war' in Ukraine since 2021.

Vladimir Putin allegedly decapitated the Wagner Group by killing the Wagner commander, Yevgeny Prigozhin, and several of his lieutenants in an aircraft accident in 2023. However, the Wagner Group and other private military companies like Redut are still on the Russian payroll. They continue to operate in Ukraine and elsewhere in Syria, Libya, the Central African Republic and Mali (Google Band of Brothers: The Wagner Group and the Russian State).

Russian capabilities in the grey zone defy easy explanation. They subvert their neighbourhood secretly through what has become known as hybrid warfare if they do not get their way in 'light grey' phases. They disguise the beginning of hybrid campaigns and inflict significant damage before they are detected, and their targeted nation can respond. They employ military, para-military and civilian agencies equipped and authorised to use violence to coerce their neighbours. Their escalations included political, military, economic, social, and informational means and conventional, irregular, sabotage, terrorism, and disruptive/criminal methods – a terrifying mix of evil for any targeted society to endure.

The Russians create 'friction' by combining seemingly random, deniable and unpredictable hostile, destructive and disruptive events, including terrorism, with concocted public protests, cyberattacks and disruption of essential services and supply chains that complicate organisational responses. Intelligence services cannot be sure where the threat is coming from. Police forces do not know who to arrest. Armies, navies and air forces cannot find other armies, navies or air forces to fight. However, the targeted population know that something is dreadfully wrong and are terrified.

Through trial and error, the Russians have invented a new Five-Dimensional (5D) mode of warfare: disinformation, destabilisation, disruption, deception, and implied destruction (Google 5GW). For Ukraine in 2021, implied destruction to coerce domination escalated to actual destruction for conquest. The CCP has watched the Russians and applied its version of grey zone suppression in Hong Kong. It is pressuring Taiwan in the grey zone continuously.

The takeaway from this brief discussion is that what might happen in Australia has already occurred in the Baltic States and Ukraine years ago and continues to this day in Russia's and China's regional neighbourhood. These campaigns are at least prototypes of what China and Russia are capable of now. Let's briefly examine some examples to set the scene for later discussion about Australia's grey zone resilience.

Estonia 2007 – subversion and economic disruption

The first known large nation cyberattacks on a smaller neighbour were the attacks on Estonia over three weeks from late April to mid-May 2007. They demonstrated how easily a hostile state can exploit simmering tensions within another society. Internal tensions in Estonia and protests in Russia arose after an Estonian Government decided to move a bronze statue of a Russian soldier built by Soviet Union (USSR) authorities in Tallinn, the Estonian capital, in 1947 amidst graves of Russian soldiers. The 'Monument to the Liberators of Tallinn' and its cemetery were prominent in the city's centre. This monument represented Russia's victory over Nazi Germany for Russian speakers in Estonia. For ethnic Estonians, it symbolised 50 years of Soviet occupation and oppression.

Nearly 75 per cent of the Estonian population voted in a referendum for independence from the disintegrating post-Cold War USSR in 1991. In 2007, the Estonian government decided to move the Bronze Soldier and the remains of Russian soldiers from the centre of Tallinn to a Russian military cemetery on the city's outskirts. Russian language media, stimulated by 'fake news' from the multiple Russian IP addresses located in Russia, disseminated a narrative that the statue, as well as all Russian war graves near its new location, were to be destroyed. The stories circulated on social media called for protests. Russian speakers in Estonia, accompanied by what an Estonian official described later as Russian-recruited 'malicious gangs', took to the streets.

On 26 April 2007, Tallinn erupted into two nights of riots and looting. One hundred fifty-six people were injured, one person died, and 1,300 protesters were arrested and detained. From 27 April, cyber-attacks tormented Estonian public and corporate institutions' computer systems for three weeks, a large-scale 'distributed denial of service'. The targets were the websites of government departments, banks, telecommunications providers and media companies. The aim was to intimidate the Estonian Government to reverse its decision to move the Bronze Soldier by crippling Estonian online infrastructure and unsettling the population. Almost 60 key websites, automated teller machines (ATM), and government e-mail stopped working. Internet trolls, 'bots,' and 'fake news' farms reinforced initial attacks. Tens of thousands of automated online requests and waves of spam swamped Estonian servers.

Electronic jamming attacks silenced Estonia media. Newspapers and broadcasters could not upload articles or broadcast the news. The only communications to the public were pro-Russian 'fake news' agitating for more protests and falsely and profanely accusing the Estonian political leaders of a range of conspiracies. Hackers disrupted Estonia's networked public and corporate institutions. There was a run on the banks after Estonians could not use their ATMs or access online banking. The riots, looting and arson damaged the city centre.

Estonia's NATO allies watched but did not intervene. The NATO Treaty's Article Five obligates members to defend each other – the basis of alliance and conventional military deterrence. But there was no act of war, invading force, or significant loss of life. Failure to identify who was responsible for the cyber-attacks made NATO retaliation problematic. They came from Russian IP addresses, and online instructions were in Russian, but there was no definite link to the Kremlin. The Russian Government denied involvement and ignored Estonian appeals for help against Russian-based hackers. The street protests and cyberattacks did not scare the Estonian Government. 'Head bowed, one fist clenched and wearing a World War Two Red Army uniform, the Bronze Soldier [now] stands solemnly in a quiet corner of a cemetery on the edge of the Estonian capital Tallinn.'

In 2017, the BBC reported, 'the Kremlin orchestrated the attack, and malicious gangs then seized the opportunity to join in and do their own bit to attack Estonia.' ... [Furthermore] Hostile states often count on copycat hackers, criminal groups and freelance political actors jumping on the bandwagon.'

The Estonian Government 'woke up'. Since 2007, Estonia has integrated government, military, and private sector capabilities for cyber defence, a new national service in the grey zone. A Cyber

Defence Unit, comprised of anonymous Ministry of Defence trained civilian IT experts, now maintains 24/7 surveillance of the cyber domain. Members of this unit donate their free time defending their country online by practising what to do if a cyberattack brings down significant utilities, such as power and water, supply chains or vital service providers, such as online banking and public transport.

Despite this example of Estonia's recovery from strategic surprise, no nation in the Western alliance appears to have a whole-of-government plan for resisting and responding to disruptive cyberattacks. NATO has established the NATO Cooperative Cyber Defence Centre of Excellence in Tallinn for regional engagement and cooperation – nice, but not enough.

Should Australia wait for a major cyberattack to establish a similar civil-military capability? Are the current arrangements sufficient to avoid strategic surprises? In Chapter 6, we argue for marshalling Australia's human capital – smart Australians – for voluntary part-time national service in the cyber domain to ensure Australia is never electronically surprised and can hit back decisively if attacked.

Georgia 2008 – cyber-attack
Russian information warfare and cyberattacks in support of a brief ground war with Georgia over a territorial dispute in South Ossetia in August 2008 was the first example of cyberattacks synchronising with a conventional military operation. The attacks were a combination of denial of service, defacing Georgian government websites, distributing malicious software, and spamming e-mail services and websites. There were no lasting impacts on the Georgian economy or political system. At best, they fell into traditional categories of sabotage, espionage and subversion, essential for an escalating grey zone campaign. Australia, take note.

Crimea 2014 – fait accompli
Russia's conflict with Ukraine before its 2021 invasion was a grey zone masterclass. It showcased the convergence of cyberattacks and contemporary military electronic warfare designed to disrupt an adversary's communications by denying them access to the

electromagnetic spectrum [EMS]. The Russian annexure of the Crimea region of Ukraine in 2014 after a *fait accompli* hybrid campaign set the benchmark and is an ominous warning.

These operations were examples of traditional electronic warfare – jamming – combined with enhanced cyber techniques developed since the 2007 Estonian attack. The Russians paralysed the Ukrainian military, law enforcement and civil command and controlled ICT systems. Spot or barrage jamming of telecommunications denied Ukrainian access to portions of or the entirety of the EMS. Russian hackers jammed drone controllers and GPS signals to bring down uncrewed aircraft and disrupt navigation systems. Ukrainian armed forces in Crimea lost their 'eyes in the sky' and technology to know where they were on the ground.

Ukrainian telecommunications networks emit unique signatures and IP addresses. Russian electronic warfare units detected these electromagnetic emissions and disrupted communications between Ukrainian vessels, land forces, aircraft, headquarters, civilian Wi-Fi, and personal cell phones. They located everyone through electronic 'signatures' and maliciously monitored, disrupted communications and manipulated social media.

Ukraine lost Crimea in 2014 and was losing the Donbas and neighbouring regions in 2021 before Vladamir Putin decided to 'roll the dice' and invade. NATO could not have mobilised in time to help Ukraine. No nation dependent on the Russians for oil and gas in the cold of winter and to power its economy wanted to pick a fight with Putin. The United States decided to fight a proxy war with its old Cold War adversary in someone else's homeland rather than closer to the US homeland. The same thinking may apply if China escalates beyond the grey zone and attacks Taiwan or another US ally in Southeast Asia or the Pacific region.

Since 2014, Russia has become the nation to watch for the scale and mischievousness of its cyberattacks. Interference with the 2016 US Presidential elections is an example. The FBI found that 'Moscow intended to harm the Clinton Campaign, tarnish an expected Clinton presidential administration, help the Trump Campaign after Trump became the presumptive Republican nominee and undermine the US democratic process.'

In 2020, one patient Russian cyberattack set a benchmark for hurting ICT-dependent governments and economies. This time, Russian hackers went global, including Australia. The SolarWinds Orion cyberattack hacked the computer systems of 18,000 public and private organisations worldwide. (Google SolarWinds Orion cyber-attack) This illustrated the capability to close public and private sector institutions, economic activity and essential services. This was a 'supply chain attack', using a trusted third-party vendor to install malware in multiple organisational networks.

Reportedly, Russian hackers had access to sensitive databases in Australia's Defence, Finance and Home Affairs departments, and the Australian Securities and Investments Commission as well as Australian Radiation Protection and Nuclear Safety Agency, the Bureau of Meteorology, trade promotion agency Austrade and the Department of Education, Skills and Employment, NSW Health, Serco Asia Pacific, and mining giant Rio Tinto for several months, developing the potential to set them up for catastrophic collapse at a future time.

The SolarWinds Orion attack supports the warning issued in January 2021 by Francis Galbally, an Australian businessman and company director involved in cyber security for 21 years:

> It is not rocket science to understand that a total attack by a rogue nation could decimate us [Australia] economically and render us instantly defenceless. ... Ransomware is perhaps the most known and prevalent form of cyberattack. It can be used for financial gain, an act of terrorism or by rogue nation-states to create havoc. In 2017, a rogue group called NotPetya smashed thousands of corporate networks around the world, costing these organisations billions of dollars. This was not about financial gain; it was proving the ability to massively disrupt global digital networks. It was originally thought to be a ransomware attack but was later traced to a Russian military unit.

China in the grey zone

The question for the Australian Ggovernment is whether the CCP could emulate Russian expertise in cyber, electronic and information warfare culminating in hybrid warfare. The CCP will not make it easy to find out. Evidence from 2019 and 2020 revealed closer collaboration between Russia and China on' high-tech'. A US National Defense University report in 2020 analysed how Xi Jinping has 'remade' the Peoples Liberation Army [Chinese armed forces] since 2015. A book published in 1999 by two Chinese colonels called *Unrestricted Warfare* informed a new doctrine of 'informationalised local warfare'. The Strategic Support Force [SSF] 'combines assorted space, cyber, and electronic warfare capabilities across the armed services and its former general departments'. The Chinese armed forces have realigned their sprawling space, cyber, and electronic warfare capabilities into one instrument. Cyberattacks are now fundamental to CCP information warfare that has a global reach. A series of high-profile cyber intrusions have demonstrated both growing sophistication and the rapid progress Chinese forces have made in a few short years.' In 2024, the SSF split into three independent arms under the Central Military Committee that represent the nature of China's future warfare: an Aerospace Force, a Cyberspace Force and an Information Support Force. (Google China's New Information Support Force)

The CCP can interfere with Australia's ICT-dependent public and private sector organisations, critical infrastructure, essential services and supply chains. Imagine the impact on ordinary Australians of interference with food supply chains to Woolworths, Coles and Aldi, power grids, hospitals, water and petrol supplies and online financial services. While we do not have space here to dissect every cyber-attack on Australia in recent years, such as the Red Ladon (TA423) group in 2022, the one against shipping container terminals in 2023 and Volt Typhoon's mischief in 2024, we conclude that the CCP is responsible for many of them (Google Red Ladon, TA423 and Volt Typhoon). They are probes and rehearsals before significant intrusions as part of an escalation from light to darker-grey tactics and possibly an electronic Pearl Harbour-like strike.

The takeaway from this section is that Australia must become capable in the cyber domain to defend against cyber and electronic attacks on ICT systems that underpin the Australian economy and way of life. The government has and continues to invest astutely, but we argue that something needs to be done to deter the CCP from attacking Australia in this domain in the first place.

How might the CCP escalate its campaign against Australia? What do we need to look for? And how do we respond?

Chinese hybrid warfare

It is improbable that Russia will prosecute a hybrid war against Australia, but could the CCP emulate Russia and prosecute its version accompanied by deniable electronic and cyberattacks? Could the CCP, frustrated by the lack of success in the light grey Phase 1, launch multi-domain political, cyber, economic, trade and information attacks? Could the Australian Government be forced to negotiate terms that compromise Australian sovereignty under pressure from an Australian political movement calling for closer relations with China in this dramatic and disrupted environment? Could all this happen while Australia's traditional allies protest but do not intervene forcefully because, like the Australian Government, they are trying to avoid a Third World War and are confused and unsure about what to do?

Readers should be sceptical of these doomsday questions. No one will ever know the mind of Xi Jinping and his CCP hardline nationalists. There is no public evidence that the CCP intends to escalate political and information warfare identified in warnings described in Chapter 1 to hybrid warfare in Australia. But the CCP has 'form'. The US Center for Strategic and Budgetary Assessments [CSBA] has analysed six CCP hybrid warfare case studies in the Indo-Pacific. It begins with the annexation of Tibet in the 1950s. It culminates with China's coercive behaviour in the Senkaku's islands in a territorial competition with Japan and its maritime grey zone campaign that has successfully established military bases in the contested South China Sea. The following detection triggers likely identify escalation from 'light grey' political and informational warfare to 'dark grey' hybrid war in Australia and the near region.

Phase 1 Escalation

- The CCP's aggressive 'Wolf Warrior diplomacy' and 'influence operations' are directed at governments, economies, institutions, and the media;
- influence operations, such as cultivation and employment of influencers, corporate and political leaders with pro-CCP narratives, and through cultural engagement, media and social media manipulation;
- The CCP's strategy includes the control of key infrastructure, such as ports, airports, telecommunications networks, media outlets, principal real estate, transport and supply hubs, islands, and remote locations;
- Infiltration of civil society is another key tactic, involving activities such as sporting and entertainment sponsorships, donations to influential charities and not-for-profit organisations, and association with culturally prominent organisations and individuals.
- infiltration of civil society, such as sporting and entertainment sponsorships, donations to influential charities and not-for-profit organisations and association with culturally prominent organisations and individuals in civil society; and
- infiltration of education systems through donations that influence curriculum and academic discourse.

Phase 2 Escalation

- the exploitation of political, religious, ethnic, regional, racial, and class differences and social cleavages;
- economic pressure through embargos and restrictive trade;
- the exploitation of criminal networks, extremists and grievance groups;
- bribery, illegal donations and other corrupt practices to gain political and civil society influence; and
- cyber-attacks and further digital disruption.

Phase 3 Escalation

- recruitment, training and employment of members of grievance groups and criminals to intimidate leaders and citizens into deterring actions that are not in the interests of the CCP or orchestrating activities in the interests of the CCP;
- the conduct of deniable and covert operations through proxies to apply violence, including terrorist acts, against institutions, groups and individuals and destroy property, accompanied by information actions, catastrophic cyberattacks and electronic warfare, and diplomatic pressure to negotiate political concessions; and finally
- The recruitment and resourcing of an Australian political party to voice the CCP agenda and convince Australians that it can end CCP pressure and deliver a peaceful and prosperous relationship with China.

The NIC, writers, and commentators have 'called out' CCP Phase 1 'light grey' persistent and competitive political warfare (see Chapter 1). The government has responded reflexively with stricter legislation and formed multi-agency civilian task forces but lacks firmer deterrence and de-escalation options for responding to further escalations to 'darker grey' hybrid warfare (see Chapter 3).

Notably, of the 14 grievances the Chinese Embassy leaked to the Australian media in November 2020 explaining why China was causing billions of dollars of damage to the Australian economy with trade tariffs and embargos, only three refer to Australia's foreign policy actions. Eleven relate to Australia's hardening of its grey zone defences with legislation and enhanced institutional machinery to counter espionage, foreign interference, cyberattacks, and foreign investment in infrastructure deemed not in the national interest. This emphasis could be a sign of the CCP's 'main game' to coerce Australia in the grey zone. It also suggests that the CCP may escalate to Phase 2 if it does not get its way in Phase 1, posing significant risks to Australia's national security.

Doomsday warnings

China will draw from Russia's 'playbook' for transitioning political warfare to hybrid warfare. The Russians deployed their special forces and intelligence agencies to recruit, arm, train, and employ grievance groups, 'patriotic' extremists, cyber militias (hackers), and criminals for their successful grey zone campaign in Crimea. They won the battle of the narratives by denying involvement while publicly supporting nationalistic Russians and those sympathetic to Russia living there.

Though Russian success in Crimea under favourable conditions may not be replicable elsewhere, the combination of military and non-military ways to annex territory is illustrative and instructive. Arguably, the CCP has already employed a variant of this combination in the South China Sea with armed fishing vessels crewed by 'patriots', described more accurately as 'Uniformed, Navy-trained fishing militia', denying access to Vietnamese, Filipino and Indonesian fishing vessels and forcefully asserting Chinese control.

There is evidence of increasing transfers of Russian military technology to China and vice versa. The war in Ukraine and Western sanctions against Russia have brought it closer to and more dependent on China. Russian hybrid warfare that is now tested and proven for escalating grey zone campaigns could inform the CCP's grey zone campaign against Australia. It is crucial to deter escalation in lighter shades of grey when tactics are non-violent, primarily political and informational. Failure to do so risks having to scramble in response to a range of non-violent and violent acts of coercion – friction – as the shades of grey darken and tactics and techniques become more violent and destructive. This underscores the need for proactive measures and preparedness.

A lot has happened in Russia and China since 2007, 2008 and 2015. Cyberattack expertise has improved, and countermeasures are in place to reduce the risk of the same effects occurring again. These campaigns are neither replicable in Australia nor exact examples of what an escalating CCP grey zone campaign might look like. Still, their characteristics serve our purpose with examples of cyber and electronic warfare-enabled pressure to coerce political concessions. They show how conventional electronic warfare and cyberattacks can combine to optimise disruption to ICT systems and what a mature hybrid warfare campaign achieved in 2014 and afterwards under the noses of NATO and the United States in Ukraine.

Taiwan's doomsday

What might an escalating grey zone campaign against Australia look like? What could be Australia's 'doomsday'? Linda Jakobson, the respected founding director and deputy chair of the Australian think tank China Matters, developed a Taiwan doomsday scenario. Her scenario, which we will discuss, is an important tool for understanding and preparing for potential CCP strategies.

Xi Jinping has no secret about incorporating Taiwan into China (Google Xi Jinping and Taiwan). Taiwan is a 'canary in the coal mine' exemplar of grey zone escalation – China's Crimea. In June 2024, Kevin Rudd, former Prime Minister and Ambassador to the United States, wrote,'

> ... we will see a change in Chinese strategy towards the 'Taiwan problem'. Indeed, we are already seeing it, with China increasingly availing itself of a 'multi-dimensional 'gray (sic) zone' strategy aimed at applying new forms of pressure on Taiwanese and international public opinion to force Taipei [Taiwanese Government] to the negotiating table. ... a 'short of war' approach – a combination of political, military, diplomatic, economic and cyber measures where the objective is to achieve a psychological, attitudinal and then behavioral (sic) change on the part of the Taiwanese public and political opinion. ... China's emerging menu of strategic measures remains short of war and short of invasion but shares the same objective: to force Taipei to capitulate.

Linda Jakobson argues that:

> A military attack on Taiwan is not the most likely route the People's Republic of China (PRC) will choose to achieve unification. Rather, the more probable scenario is *a strategy of 'all means short of war'* [authors' emphasis] in which the PRC would attempt to force Taiwan to the negotiation table through a mix of pressure tactics, including military intimidation, dissemination operations, cyberattacks and covert actions [hybrid warfare]. The United States and others, including Australia, would find it extremely hard to counter these moves. No individual action by the PRC would warrant a military response. ... In an

attempt to break the will of Taiwan, Beijing could adopt an aggressive mix of new technologies and conventional methods to apply pressure. These range from economic pressure or an embargo via intimidation, cyberattacks, covert actions and subversion to assassination and the limited use of military force.

While acknowledging that the relationship between China and Taiwan and China and Australia are not comparable and have no geographic, historical or cultural similarities, Taiwan's doomsday scenario does foreshadow Australia's doomsday. Linda Jakobson provides a chilling scenario to ponder. The CCP's political objective for Taiwan is unification and acceptance of 'One China'. Its political objective for Australia is to be a contemporary Chinese tribute state bound by a 'two systems' agreement for peace and prosperity. Could ways and means for incorporating Taiwan be employed to coerce an Australian government to negotiate for 'tribute state' status, presented as 'a partnership for peace and prosperity'? Linda Jakobson hypothesises that a CCP Phase 3 escalation in Taiwan might look like this:

Scenario: 'All means short of war.'
In this scenario, the PRC would not invade Taiwan. Rather, Beijing would strive to create utter chaos in Taiwan and compel the government to accede to the PRC's demands. Initially, it would be impossible to credibly pinpoint who is behind many of the provocative actions. Few shots would be fired other than for possible political assassinations. Taiwan's armed forces would struggle to counter Beijing's actions. Barring strong condemnation of Beijing and imposing economic sanctions on the PRC, the US and others, including Australia, would find it difficult to assist Taiwan.

This scenario could start with PRC officials gathering major Taiwanese investors in the PRC and insisting that they sign a letter to Taiwan's government calling for cross-Strait political talks. Refusal to sign would result in business difficulties. Xi Jinping would also urge Taiwan's 'leader' [President] to immediately agree to consultations to collaboratively seek unification.

Next, Beijing would suddenly cut Taiwan's air routes into PRC cities, stating that foreign airlines needed those routes. International airlines would be told to choose between flying to the PRC or Taiwan. PRC combat aircraft would conduct incursions not only across the median line of the Taiwan Strait, as they do today, but over Taiwan itself. Would Taiwan's Air Force be directed to shoot down such intruders and risk an all-out war?

Taiwan's stock market would presumably plunge. In this situation, the Democratic Progressive Party, the current ruling party that leans toward independence, would encourage legislators to insist on 'no preconditions for political talks'. PRC-backed media outlets in Taiwan would run scare campaigns. Protesters would take to the streets. Some groups would demand a declaration of independence; others would demand that the government open political talks with Beijing. Street gangs would attack independence supporters. Confrontations between opposing political groups could become violent. [Note: The CCP employed this technique in Hong Kong.]

The campaign's most intense phase would include the PRC ramping up disinformation efforts and launching a barrage of sophisticated cyberattacks with the aim of first disrupting Taiwan's electricity and telecommunications and then shutting them down. ...

Rumours of the PRC's intentions would run rampant through Taiwan's darkened cities cut off from communications. The PLAN [Chinese navy] would start operations to impose a partial blockade of Taiwan's western harbours. Beijing would request governments to shut down their representative offices in Taipei [capital of Taiwan]. An editorial in the People's Daily would encourage Chinese compatriots in Taiwan to make the right decision, warning that the clock is ticking.

Australia's doomsday

Surely applying this Taiwanese doomsday scenario to Australia is far-fetched and ridiculous? Isn't it preposterous that the CCP would find a group of Australian corporate leaders connected to trade with China to pressure the Australian Government to negotiate a partnership for peace and prosperity? The CCP would not impose economic pressure on Australia, including closing Chinese ports and airports to Australian vessels and aircraft. Intermittent disruption of telecommunications and Internet access would not unsettle the population. The Australian people would remain strong and stoic in the face of state-of-art cyberattacks on Australia's critical infrastructure (energy, water, fuel and food supply) that coincided with a social media campaign warning that all bank accounts had been hacked and funds were not accessible.

There would be no panic or antisocial contagion behaviour in Australia. Stampedes for toilet paper during the COVID-19 pandemic were one-off behaviour. Australians would take the disruption of power, food, petrol, water, and electronic transactions in their stride and trust the government to respond efficiently and effectively. To think otherwise is insulting and un-Australian!

We're sorry if you take offence at our scepticism, which borders on sarcasm, and what follows. We believe what we have argued in Chapter 1. We believe that the NIC, Kilcullen, Connolly, Hamilton, Joske, Fitzgerald, Hartcher, Brady, the US CSBA, and others are right.

Let's see what a 'perfect grey zone storm' in Australia might look like. Its purpose is to coerce Australia, a significant source of mineral resources and agricultural produce, a nation with substantial political influence in Southeast Asia, and a middle-power ally of the United States, to act in China's national interests by signing an agreement for peace and prosperity. Some of the events described in this fictional scenario occurred in 2014–2024.

Phase 1 might be CCP United Work Front operatives and their proxies infiltrating political institutions, corporations, universities, and civil society in Australia and its regional neighbourhood. They intend to garner political influence and co-opt or compel individuals and groups. Phase 1 aims to undermine and shape the political and economic order and public narratives to favour CCP interests. A dominant narrative is that it is in Australia's national interest to establish a partnership for peace and prosperity with China. Anyone in Australia opposed to China's global aspirations endangers a productive, mutually beneficial trading relationship. Indeed, it would become un-Australian to question Chinese motives. The same tactics would apply in the Pacific Islands and Timor Leste, where it would be un-patriotic to criticise China and interfere with lucrative development projects.

A well-funded coalition of ex-politicians, corporate leaders, societal influencers, journalists and commentators provide a public platform for CCP narratives. This coalition forms an anti-American and pro-China Forum for Peace and Prosperity [FPP]. It promotes the idea of American strategic decline and mocks its political and social divides. The FPP calls for Taiwan to join the Chinese motherland, claiming to have unique leverage in China. It seeks credit for opening new business opportunities to demonstrate that China wishes to partner with Australia without pressure. Xi Jinping and senior CCP leaders visit Australia to offer mutually beneficial trade and security agreements and promise a future of peace and prosperity. There are optimistic affirmations of the mutual benefit of a strong China-Australia relationship based on trade in public, like the G20.

In response to the Australian Government's assertion of sovereignty, affirmation of the US-Australian alliance, and irritation with AUKUS, the lease and purchase of nuclear-powered submarines from the UK and the US, and the QUAD, a collaborative dialogue on regional security between Australia, India, Japan and the United States, the CCP escalates to Phase 2. More severe cyberattacks occur on shipping container terminals, banks, power grids, supply chains, and public and private sector institutions and organisations, undermining public confidence in the Australian government and its security arrangements.

The CCP denies involvement in these attacks but increases economic embargos and other financial pressures. Behind the scenes, penetration of major Australian political parties continues apace. The intention is to create an uncertain environment to precipitate a State or federal government change to one more favourable to CCP interests. There is increasing evidence of Chinese hackers manipulating Australian electoral and opinion poll processes manually and electronically.

The FPP facilitates operatives and sympathisers of Chinese heritage among the Australian community and steps up its campaign to radicalise others to support China's plan for a partnership for peace and prosperity. Protests begin. They support China, are anti-American and condemn the Australian Government's legislative actions to thwart the CCP grey zone campaign. These protests prompt counter-protests from White Supremacist and Far-Right activists, calling on patriotic

Australians to stand up to Chinese bullying and warn of 'invasion'. There is evidence that CCP agents are provoking these groups to act violently. Police struggle to stop physical violence between protest groups.

Political pressure on the Federal government increases to appease China and negotiate to end civil unrest. The FPP amplifies United Front Work narratives, suggesting the US alliance is unreliable and obsolete considering China's inevitable rise and entitlement to influence in the Asia-Pacific Region. Concocted civil protests occur in cities in the Pacific Islands, and Timor Leste calls for closer relations with China. Chinese-trained police forces stand by while protests become more violent. New Zealand calls on Australia to recognise China's ascendency while boasting of its capture of Australian market share in the Chinese economy.

Phase 3 escalation begins with several acts of terror, including bombing headquarters of law enforcement and intelligence agencies. Australia's law enforcement capabilities succumb to special operations. In short order, Australia's counter-terrorism hunters become the hunted. Deniable terrorist acts and other disruptive acts of coercion and provoked public disorder escalate to a range of deniable coercive actions through proxies associated with the prosecution of hybrid war. The FPP blames the government for causing the deteriorating situation and calls for an election to allow the people to decide on better policies for forging a new relationship with China. Opposition party factions call for a radical new approach to secure Australia's prosperous and peaceful future in partnership with China.

Australia faces the Phase 3 escalation alone despite participating in every 'coalition of the willing' since the end of the Cold War. There is no US or UN-sponsored 'coalition of the willing' to help Australia counter a hybrid attack disguised as escalating political differences among Australians over managing Australia's relations with China. The ANZUS Treaty is irrelevant because there is no military attack. There is no invasion fleet on the horizon: no navy, army or air force to fight. The government is left to call out the ADF to protect critical infrastructure and stand by if civil unrest overwhelms law enforcement agencies. President Trump authorises the United States to assist Australia with intelligence support and arms. Other American leaders join Britain in public criticisms of the CCP and promise further help if things worsen. The United Nations Security Council is impotent – China vetos resolutions condemning their perceived involvement in destabilising Australia.

In an unprecedented surprise move, the government and the opposition split internally over Australia-China relations. A rump of elected opposition members and a faction from the government affirms that Australia must negotiate an agreement with China or risk economic and societal collapse. They form a new coalition called the Party for Peace and Prosperity (PPP), calling for a snap election to decide on Australia's relations with China. They claim that their electorates have had enough, and that ordinary Australians have spoken to them. Elected politicians from the government, the opposition and minor parties 'cross the floor' of the Parliament to join the PPP. The two-party system collapses. After a turbulent meeting of the Federal Executive Council, the Governor-General-in-Council exercises prerogative powers and authorises a dissolution of Parliament and an election under Section 28 of the Constitution.

In the most violent election in Australia's history, with claims and counterclaims of foreign interference in the form of cyber-attacks and concocted civil unrest, the PPP, made up of politicians from all political parties and supported by companies and organisations doing business with China, is elected. Suddenly, all violence ceases. China announces a return to the terms of free trade agreements and promises an enhanced agreement on security favourable to both countries. Statements and speeches affirm Australia's sovereignty and welcome Australia to the Belt and Road Initiative and the new world order of peace and prosperity.

Conclusion

This chapter cannot predict the future. Still, it is foolish to ignore the past and suggest that the CCP will not act on its publicly stated intentions or repeat past actions in the future. While the Russians are the grey zone maestros, the CCP are the virtuosi, as the citizens of Tibet, Taiwan, Hong Kong, and the Philippines can attest. President

Trump plans to supersede the rules-based global order with a deals-based order that avoids a Third World War. While the world will be grateful to prevent another global war that might risk the nightmare of nuclear war, the axis of autocrats will use the grey zone to further their interests.

Many readers may find the Australian grey zone doomsday scenario unthinkable, ridiculous, provocative and unpatriotic. Still, countering political and hybrid warfare is Australia's sovereign responsibility, whatever the deals-based future may hold. We should neither assume assistance from allies nor rely on asking them to overcome Australia's lack of preparedness in the grey zone. Australia should signal to friend and foe alike and its regional neighbours that it comprehends the CCP grey zone campaign and has developed lawful and carefully calibrated strategic responses. Australia is well prepared to defend its sovereignty and vital interests firmly and independently. Australia is also ready to partner with regional neighbours if they admire Australia's new fair but firm 'Let's trade, not argue' approach. They would appreciate support with their de-escalation strategies.

Before laying out the book's de-escalation strategy and the ways and means to implement it, let's deal with the belief that the Australian government knows best, is already taking sufficient action, and will guide Australia through these troubled times: 'She'll be right, mate!'

CHAPTER 3

She'll be right, mate?

Introduction

This chapter is for optimistic readers who believe the government will get it right. It explores whether responses to the new Cold War and preparing Australia in the grey zone are adequate. After all, Federal governments are responsible for keeping Australians safe and have done a great job since the end of the first Cold War, so 'She'll be right, mate.' Maybe not.

History may not repeat, but the 2020s rhyme with the 1910s and the 1930s when liberal democracies didn't deter bullying. War engulfed the world twice and cost millions of lives. The Australian Pub Test – common sense – applies – if you don't stand up to your bullies, they won't stop. Deterrence does not work unless a bully fears consequences. Australia isn't ready to deter CCP bullying. Relying on others to protect Australia from grey zone coercion risks freedoms and sovereignty and insults Australian nationhood. The potential risks of this reliance are significant and should not be underestimated.

Australia's 2024 National Defence Strategy (NDS2024) rhymes with Australia's choices in the 1910s and 1930s: dependence on a powerful ally and overdue investment in military hardware. It was then ships and aircraft; now it's submarines and missiles. The NDS2024 correctly assesses that 'Australia faces the most complex and challenging strategic environment since the Second World War. It demands a coordinated, whole-of-government and whole-of-nation approach to Australia's defence.' The NDS2024 aims 'to ensure that Australia becomes more capable, self-reliant and takes greater responsibility for its own security.'

The NDS2024 puts the physical defence of Australian territory from invasion first. The map that guided its authors positioned Australia southeast of islands stretching to the northwest and north to China –

an invasion's stepping stones. Its bedrock idea is that Australia must have armed forces that can project force to the northwest and north to defeat an enemy using these stepping stones as forward bases for its navy, army and air force. The NDS2024 investment in a denial strategy of 'enhanced lethality and greater range' focussed on deterring an invasion through the air-sea gap and islands to Australia's north is Australia's primary and most important defence strategy. This book does not question this approach.

The NDS2024 priority for territorial defence is fair enough if China chooses to roll the Japanese strategic dice and repeat the 1941–1945 Pacific War. No substantial evidence exists that China has the will or military weight to seek domination through territorial conquest. The proof of a grey zone campaign to achieve dominance as part of an indirect approach to winning without fighting a conventional war is abundant and in plain sight (See Chapter 1).

The CCP grey zone campaign bypasses geographic stepping stones to destabilise the Australian homeland. NDS2024 does not address the 'grey zone activities' it identifies or deal with 'the prospect of coercion' in the grey zone it warns about. It only glimpses grey zone tactics that ignore geography, such as cyber-attacks that can disrupt critical essential services, such as energy, transport, water, sanitation and medical services, infrastructure, communication networks, and supply chains, such as food, fuel and other commodities. There is no mention of deterring CCP espionage, disinformation, political interference, or influence operations.

This book proposes a grey zone strategy that hedges and enhances Australia's NDS2024 strategic bets on the US alliance, nuclear-powered submarines, and longer-range missiles. It is primarily a diplomatic deterrence strategy that proposes innovative and affordable ways and means to keep Australians safe and prevent war. The message to the CCP is, 'Let's trade, not argue, but if you bully us, there are uncomfortable consequences.' The uncomfortable consequences escalate in response to the CCP moving from 'light grey' political and economic pressure to 'darker grey' coercion and disruption.

The strategy draws guidance and lessons from Australia's Pacific War experience when the nation last faced an Asian power determined to dominate Australia's near region across its trade routes to the rest of the world. It optimises Australian human capital and leverages mutually beneficial trade, Australia's most significant strategic advantage after geography. The aim is prevention, not provocation. If war looms, Australia will again appreciate help from allies, but unlike in 1942, Australians will have a sovereign strategic option. Sovereign means a self-reliant strategy Australia can implement independently with its resources to keep Australians and their homeland safe.

Since the beginning of the CCP campaign in 2014, Australian governments have acknowledged the grey zone and taken some action. However, strategic guidance and the ways and means to counter an escalating CCP campaign are still not joined up or sufficient. The 2016 Defence White Paper is silent. The Defence 2020 Strategic Update mentions 'grey zone activities' for the first time but does not offer a strategy. While acknowledging the cyber and space domains, it recommends buying more ships, submarines, strike aircraft and longer-range missile systems, almost as an extension of the 2016 Defence White Paper. The 2023 Defence Strategic Review (DSR2023) also mentions the grey zone in passing before calling for massive expenditures on ships, submarines, missiles, etc.

There is bipartisan support for the promise of the 2021 AUKUS Agreement that deepens Australia's relationships with traditional allies, the United States and Britain, this time for nuclear-powered submarines by 2040. The NDS2024, based on DSR2023, confirms that Australia's defence strategy is to rely on the United States and buy more firepower. The book does not argue against AUKUS and Australia's other measures to keep Australia safe. AUKUS is an excellent deal for access to American and British technology alone. Our point is that Australian strategic guidance and initiatives do not adequately address the grey zone.

The Department of Foreign Affairs and Trade (hereafter Foreign Affairs) identifies the grey zone threat but has no strategy to counter it. In February 2024, Penny Wong, Australia's foreign minister, warned that:

Our security in the region is challenged by actions that fall far short of conflict [grey zone activities]. Throughout the Indo-Pacific, there is an urgent need to address disinformation, interference, opaque lending practices and coercive trade measures. ... This all encroaches on the ability of countries to act in pursuit of their interests.

Her department joins Defence in not offering a strategy for countering an escalation of these 'actions that fall far short of conflict'.

In this chapter, we acknowledge what the government has done and the foundations that exist to get Australia ready in the grey zone, but we argue that 'She won't be right, mate'. Australia urgently needs a de-escalation strategy and the ways and means to implement it in the next five years. But it is not all bad news. The Abbott, Turnbull and Morrison governments responded astutely to homeland threats, such as terrorism and illegal asylum seeker boat arrivals, with new homeland security arrangements and legislation. The book's strategy is built on these foundations.

After the CCP escalated its grey zone campaign with increased espionage, political interference, aggressive diplomacy and trade embargoes in 2015–16, the Turnbull Government introduced legislation that passed into law under the Morrison Government by 2018–19 before the COVID pandemic hit and 'paused' everything. Some excellent organisational and legislative foundations exist to build capabilities to counter the CCP grey zone campaign in the homeland. Let's briefly review them to give some comfort, but not enough assurance that 'She'll be right, mate.'

Homeland defence – the grey zone frontline

The Department of Home Affairs (hereafter Home Affairs), established in 2017, is a good fit for the domestic grey zone. This department aligns responsibilities, authority, accountability, and resources with the Commonwealth's role in law enforcement, border security, counterterrorism, cyber-security, and social cohesion. The Home Affairs portfolio now brings together Australia's federal law enforcement, national transport security, criminal justice, emergency management, multicultural affairs, settlement services, and immigration and border-related functions to keep Australia safe. The Minister for Home

Affairs has the tools to secure the homeland from harm, coordinate emergency responses to hazards like bushfires and floods, and counter external threats such as irregular migration and transnational crime. Home Affairs can adapt to counter the CCP grey zone campaign at home, including establishing and employing the national component of a Response Force – more on Response Force later.

The National Intelligence Community (NIC) is Australia's detection tool in the grey zone. The Office of National Intelligence [ONI] advises the Prime Minister and the National Security Committee of Cabinet (NSC) on matters of strategic importance. ONI also manages the intelligence enterprise, coordinates and evaluates Australia's foreign intelligence activities, and leads the NIC, comprised of ten intelligence gathering agencies (Google 'The Australian National Intelligence Community'). The NIC's domestic flagship is the Australian Security and Intelligence Organisation (ASIO). The Australian Federal Police [AFP] is the instrument for enforcing homeland security legislation. So far, so good.

Responses so far

Australia has responded to the CCP Phase 1 'light-grey' escalation. In 2015, John Garnaut reported this escalation to the Turnbull Government. Australia hardened legislation and established task forces against espionage, foreign interference, dodgy investments in Australian infrastructure, and resilience against cyberattacks. To be clear, CCP espionage steals defence, political, industrial, foreign relations, commercial or other information for China's advantage. Espionage became a crime in Australia, punishable by up to 25 years imprisonment, a helpful deterrent. CCP foreign interference is coercive, corrupting, deceptive, clandestine (meaning undercover), and is contrary to Australia's sovereignty, values, and national interests. According to the 2017 Independent Intelligence Review, foreign interference is strategic, economic, societal and political:

> By wielding undue influence on the Australian political landscape, foreign adversaries have the potential to undermine Australia's sovereignty [independence] and system of government. ... [Foreign

interference] can cause severe harm to Australia's national security, compromising Australia's military capabilities and alliance relationships, and can pose a grave threat to Australia's economic stability and well-being.

Espionage and foreign interference legislation give Australia new grey zone tools. In December 2019, Prime Minister Scott Morrison announced the establishment of a new Counter Foreign Interference Taskforce in Home Affairs that builds on an earlier investment in a Foreign Interference Threat Assessment Centre in ASIO. The new task force aims to 'discover, track and disrupt foreign interference in Australia.' These task forces and a counter-foreign interference strategy are astute responses to Phase 1 'light grey' CCP political warfare. ASIO now detects espionage and foreign interference, and the AFP detains, deports or imprisons perpetrators. The first grey zone combatant, Di Sanh 'Sunny' Duong, went to jail in 2024 for attempting to influence Liberal minister and Prime Ministerial aspirant Alan Tudge (Google Di Sanh 'Sunny' Duong). So far, so good.

In his 2024 Threat Assessment, Mike Burgess was specific:

When we see more Australians being targeted for espionage and foreign interference than ever before, we have a responsibility to call it out. Australians need to know that the threat is real. The threat is now. And the threat is deeper and broader than you might think.

... Right now, there is a particular team in a particular foreign intelligence service focusing on Australia – we are its priority target. ... In a sign of how the threat [from this team] has grown, successful disruptions [against this team] have increased by 265 per cent and continue to increase exponentially [escalate].

In a sensational announcement, Burgess called out an unnamed former senior Australian politician who:

sold out their country, party and former colleagues to advance the interests of a foreign regime. At one point, the former politician even proposed bringing a Prime Minister's family member into the spies' orbit. Fortunately, that plot did not go ahead, but other schemes did.'

Let's be clear: Burgess is calling out the CCP.

What about cyber-attacks?

Responding to the 2015 escalation, the Turnbull and Morrison governments hardened Australia against cyberattacks. Disruptive CCP cyberattacks on public and private sector institutions prompted the government to establish the Australian Cyber Security Centre within the Australian Signals Directorate [ASD] as well as Cyber Command within the ADF in 2018, and for the nation's counter-cyber warfare policy-making and coordination to find a place in the Home Affairs portfolio. Though not publicly acknowledged, CCP cyberattacks also prompted the government to place the Australian Cyber Security Centre under ASD's direction and separate the ASD from Defence as a statutory authority to make it a national asset, a helpful precedent when considering establishing a Response Force in the following chapters. These new arrangements did not deter a so-called 'sophisticated state actor' from conducting a major cyber attack on public and private sector institutions on 19 June 2020.

In response to the 19 June 2020 attacks, Scott Morrison announced the 2020 Cyber Security Strategy as 'the largest ever Australian Government financial commitment to cybersecurity'. He stated:

> The Government will introduce legislation to bolster the powers of the Australian Federal Police and Australian Criminal Intelligence Commission to identify individuals and their networks engaging in serious criminal activity on the dark web. Powers that allow offensive disruption capabilities will allow [Australian] law enforcement to take the fight to the digital front door of those using anonymising technology for evil purposes. ... Agencies will be equipped to help address sophisticated threats [implying the CCP], *mainly the essential services Australians rely on – everything from electricity and water to healthcare and groceries* [author's emphasis].

By 2024, ASD was issuing 'Alerts and Advisories' online regularly about the latest cyber threats. On 20 March 2024, ASD identified the CCP-sponsored Volt Typhoon as pre-positioning in US infrastructure, implying a similar tactic in Australia. We appreciate warnings but argue that the CCP must be deterred from cyberstalking and cyber-attack rehearsals in the first place.

Joining up

We welcome initiatives to counter espionage, foreign interference and cyberattacks, but governments struggle with strategic integration. Defence and other departmental grey zone strategies are not joined up. There is an urgent need for a capstone National Security Strategy – an NSS and an NDS – that includes guidance and arrangements for countering the CCP grey zone campaign. Defence's NDS2024 focuses on defending Australia's air-sea gap to the north and northwest with submarines, surface vessels, missiles and aircraft. It is disconnected from Home Affairs 'sub-strategies' to counter grey zone threats that include the Cyber Security Strategy (2020), Countering Foreign Interference Strategy (2020), Australian Disaster Preparedness Framework (2018), Critical Infrastructure Resilience Strategy (2015), and the Australian Counterterrorism Strategy (2015).

No single strategy includes the grey zone in an 'all hazards' and 'all threat' approach to keeping Australia safe. This comprehensive strategy would anticipate future hazards and give Australia the tools to deal simultaneously with the consequences of climate change and threats such as grey zone campaigns that can distress and displace Australian communities in the homeland and escalate to violent conflict.

Observations

The CCP engaged Australia's homeland defences in 2015–16, and the Australian Government responded. Still, the challenge for the Australian Government is moving from reacting to the CCP's next move to anticipating escalation and having well-rehearsed de-escalation capabilities. The NIC can detect, ASIO can disrupt, the AFP can detain and deport, and the courts can jail grey zone combatants to create some domestic deterrence. But these are reflexive responses to 'light grey' escalation, not a strategy anticipating and deterring 'darker grey' national and international escalation.

CCP hardliners are counting on the Australian Government to remain one step behind in optimising the chance of strategic surprise and success. A better option is to commit to a strategy that informs protective legislation that hardens and rehearses Home Affairs, the NIC and AFP to pace and deter further escalation and intensify responses if deterrence fails. Let's not give the CCP the first grey zone punches every time.

The critical point about deterrence is that hurtful responses to grey-zone bullying must be real and ready. The 2020 Defence Strategic Update mentions the connection between ADF special forces and countering grey zone activities in the future. Legislation and institutional machinery to deter escalation are needed now, not in the future. The Australian Government cannot afford to wait for a destructive or fatal incident or a catastrophic cyberattack to respond forcefully. At best, arrangements for ADF special forces to respond in the grey zone are ad hoc and do not systematically anticipate complex and enduring escalation with well-rehearsed contingency operations.

The success of deterrence depends on adversaries understanding the risks they will take if they are detected planning for or executing acts against Australia's national interests. They must inspect legislation that targets them and comprehend the hurtfulness of responses available to the Australian Government. An analogy would be householders putting a sign on an entrance gate, 'Beware of the dog that we are planning to buy' rather than 'Beware of the dog'. Australia's grey zone guard dog is this book's Let's Trade, not Argue Strategy, backed by a well-rehearsed Response Force.

In the past, emergency wartime legislation and substantial preparations have not occurred without a declaration of war or public agreement that war was imminent. Geoffrey Blainey, an eminent Australian historian, wrote about Australia's preparedness for a war on the 75th anniversary of the end of the War in the Pacific in 2020, hinting at the CCP's grey zone campaign:

> Nevertheless, one thing is clear: democracies can often display
> dedication in fighting a war once a war occurs but still fall *dangerously*

short when the time comes to prepare for tensions or *even for a cold war* [author's emphasis]

What about the Department of Defence?

While Home Affairs, new national security legislation, foreign interference and cyber taskforces cover Phase 1 'light grey' escalation and the NIC and ASIO can detect escalations, these responses do not de-escalate hybrid warfare. Surely, Defence is the 'go to' department to anticipate and counter this armed and destructive 'darker-grey' escalation? Defence is the wrong hammer for the grey zone nail unless war is imminent. Figuratively, Defence will mobilise when 'the gloves are coming off', and everyone knows that the 'blue' is on.

Defence is the wrong hammer because it's preparing to hit other nails. Each armed service (Navy, Army, Air Force) understandably competes for resources to prosecute a war against the traditional threat of invasion, not to counter 'dark grey' escalation in the homeland. The Navy prepares to fight another navy. The Army prepares to fight another army. The Air Force prepares to fight another air force. That is Defence's core business. But could Defence adapt to the grey zone? Could the government use the Defence hammer to hit the grey zone nail? The answer is 'No'. A fourth armed service, called Response Force and explained in Part 3, is the proper hammer.

Without clear Prime Ministerial and Cabinet direction, Defence cannot respond independently to internal threats to Australia's security that have traditionally been the responsibility of other Federal departments and agencies as well as State and Territory law enforcement agencies. Espionage, political warfare, information warfare, influence operations, cyber-attacks, and other coercive tactics elevate and complicate threats for the Home Affairs and Attorney General's portfolios, not Defence. The chances of the Prime Minister and Cabinet directing Defence to get Australia ready in the grey zone are negligible – Buckley's.

Defence is not clueless, but the department has no incentives or jurisdiction to get Australia ready in the grey zone. The 2020 Strategic Update promised grey zone capability development:

[Australia needs to develop] capabilities to hold adversary forces and infrastructure at risk further from Australia, such as longer-range strike weapons, cyber capabilities and area denial systems [Navy and Air Force]. ... These capabilities must deliver deterrent effects against a broad range of threats, including *preventing coercive or grey-zone activities from escalating to conventional conflict.* [author's emphasis] ... The new policy will require force structure and capability adjustments focussing on responding to grey-zone challenges, the possibility of high-intensity conflict and domestic crises.

The Update cites ADF special forces as the option for countering grey-zone activities, emphasising that they are 'increasingly important to countering the grey-zone threats Australia is likely to face in the future.' The Update emphasises information, cyberspace, and space as new domains for warfare. There is a promise of 'increased investment in capabilities to respond to grey-zone activities, including improved situational awareness, cyber capabilities, electronic warfare and information operations.'

Commendably, there is also a focus on growing 'the ADF's self-reliance for delivering deterrent effects.' However, this additional deterrence is primarily related to longer-range missile systems. Still, the 2020 Update mentions expanding 'Defence's capability to respond to grey-zone activities, working closely with other arms of Government.' – a promise but not yet a strategy, and there are no signs of Defence getting ready in the grey zone since the Update. We apologise for our scepticism if substantial and astute preparations have been made secretly.

The Update does enhance traditional military capabilities. Scott Morrison announced upgrades to Australia's Navy, Army and Air Force and reminded the nation of the 1930s and the collapse of strategic warning time from 10 years to zero. He focussed on Australia's immediate region and notified regional adversaries that Australia would project military hard power further from the homeland if needed. He emphasised:

We need to hold our potential adversaries to a greater distance. Part of our repositioning is to hold them further away and to work with multiple partners to achieve our goals of regional stability, peace and security. ... These new objectives [shape, deter, respond] will guide all aspects of Defence's planning, including force structure planning, force generation, international engagement and operations.

The new concepts of 'shape, deter, and respond' add sophistication and flexibility to Australia's traditional geographically focused military strategic guidance, i.e., defending the air-sea gap around Australia. Shaping is obvious, and deterrence is commendable because it focuses on having and communicating credible and forceful consequences for bullying or committing violent acts against Australia.

We will discuss extended deterrence from the US alliance later, but let's put to rest any notion that the ADF will deter CCP regional tribute state ambitions. We agree with Professor John Blaxland, a fellow soldier-scholar, who offers the following sobering and realistic assessment of Australia's conventional deterrence over the next 10 to 15 years, the lifespan of Xi Jinping's rule in China:

> ... while the ADF is a capable force, should Australia ever face a challenge from a nation with advanced weapons systems, this force may be inadequate for the task. A one-division regular-army force of three combat brigades and some special forces, a navy of a dozen or so warships and a handful of submarines, and an air force of only 100 fighter aircraft means Australia has little if any ability to sustain significant attrition in case of a substantial conflict. *In effect, the ADF is only a one-punch force* [authors' emphasis].

Justin Bassi, the ASPI Executive Director, wrote mournfully in the Foreword of the annual 2024 Cost of Defence Report, '

> ... we are in the gravest geopolitical period in generations, and this is only going to intensify. ... In particular, this year's [Defence] budget priorities are not directed towards strengthening the ADF's ability to fight in the next decade. ... If war were to break out anytime in the next ten years, our military [ADF] would essentially fight with the force it

has today. Based on current resourcing, nothing significant will change over the [next] decade.

However, it is not just the 'weight' and 'one punch' of the ADF that makes it irrelevant to deterrence; it is the 'will' of Defence and the ADF to adapt. Defence public servants, the Defence industry and ADF officers favour firepower over a complementary de-escalation strategy. They want more and better of the same for the Navy, Army and Air Force to defend the air-sea gap and to 'stick to the knitting' of depending on the United States for Australia's defence. Also missing from Blaxland's 'one punch' assessment is how many punches the ADF could absorb and remain standing after it threw its first and only punch.

Australia's updated 'shape, deter and respond' strategy is insufficient. It implies conventional retaliation if deterrence fails and is unprepared for further mobilisation. Arguably, Australia will never be able to afford the projection of sufficient quote 'credible military force' unquote to deter a significant Asian military power like China in a more contested Indo-Pacific region. American commentators Matisek and Bertram opine that conventional military power is 'practically irrelevant in stopping the attainment of political objectives by Russia, China, and others [in] 'gray-zone conflicts.' Alan Dupont, an astute Australian soldier-scholar, offers that 'plausibly deniable operations', use of 'proxies to help achieve strategic objectives', and 'manipulation of our democratic processes and institutions for political gain' … 'makes it politically difficult for democracies to respond militarily.'

David Kilcullen points to Qiao Liang and Wang Xiangsui's 'side principal rule' explicitly in their blueprint for Chinese future warfare, *Unrestricted Warfare*. In the context of this chapter, this rule suggests that Australia may be exhausting itself in a 'comfort zone' of massive investment in nuclear-powered submarines, frigates, missiles and fighter aircraft. This preference and a traditional reliance on allies distract from the possibility of a CCP 'unconventional side stroke from an unexpected direction', such as political and hybrid warfare supported by an electronic cyber-Pearl Harbour. Conventional deterrence from an enhanced ADF is necessary, comforting and politically responsible, but a grey zone de-escalation strategy should accompany it.

Ultimately, the Defence 2020 Update, DSR2023 and NDS2024 task the ADF to defend the air-sea gap and deny the invasion stepping stones. Identifying the grey zone as a future threat is commendable but is not a strategy, and there is no evident capability development. We apologise if there is – possibly in the classified version of NDS2024 – we hope so. The CCP is adopting an evasion rather than an invasion strategy. Its grey zone campaign ignores the air-sea gap and is carefully calibrated to avoid military retaliation.

Won't submarines be enough?

Our billion-dollar question – or should it be a trillion-dollar question? – Will AUKUS submarines deter the CCP from escalating Asia-Pacific tensions in the late 2020s and mid-early 2030s? Xi Jinping is working hard to achieve China's global destiny in his lifetime. He has 10 to 15 years to achieve his goal before his 80s. Inconveniently, AUKUS submarines are decades away from delivery. However, there are proposals to lease three US Virginia Class nuclear submarines in the 2030s until the AUKUS submarines come online in the 2040s and 2050s.

A famous Australian strategist, Hugh White, likes submarines, but not AUKUS ones (Google Hugh White Dead in the Water). His books, *The China Choice, How to Defend Australia, Without America: Australia in the New Asia* and the essay *Sleepwalk to War* contend that China's rise to a military superpower is 'a given' and America's decline in the Asia-Pacific region is inevitable. Australia should not rely solely on the United States to defend Australia's national interests in a more contested Asia. We agree with him so far.

White makes a case for Australian self-reliance. We support that aspiration. He proposes a 'defence from invasion' grand strategy through 'sea denial', calling for a 32-strong submarine fleet and more strike aircraft, and consigns the Army to defend the homeland and engage in guerrilla warfare after an invasion if the defence of the air-sea gap and the invasion stepping stones fails. He assigns the Air Force to support the Navy in attacking an inbound invasion armada in the Indonesian-Melanesian archipelago, presumably after achieving air superiority – Australia's future battle of Midway, probably without the support of the US Navy and Air Force.

White is silent about where the money will come from for 32 submarines, where and how they might be built and who will run them. He is also silent about political and hybrid warfare in the grey zone. He dismisses cyber warfare as neither a decisive strategy nor a viable warfare domain because of its mutually assured destructive (MAD) effects. We guess he means mutual economic suicide will deter the CCP from executing a surprise electronic Pearl Harbour and the United States responding, 'in kind'. Australia does not have sufficient counter-cyber capabilities to create White's MAD scenario with China. In 2024, Australia's posture remains primarily defensive for cyber warfare.

White dismisses the AUKUS submarine promise, joining Sam Roggeveen, a fellow academic, and others in 'sinking' the idea in favour of an independent Australian strategy. Roggeveen ignores White's 32-submarine invasion fleet ambush in favour of an 'echidna strategy', making Australia a more difficult target for China. The metaphorical echidna's 'spikes' are missiles, mines, cyber weapons and smaller naval vessels while maintaining current ADF capabilities. White's call for 32 submarines is ambitious. But metaphorically, White's archipelagic submarine ambush and Roggeveen's spiky echidna are not viable hammers for the grey zone nail.

White and Roggeveen are silent about what Australia should do to deter the CCP in the next five years. Scott Morrison was wise to remind Australians that the 2020s are like the pre-war 1930s and Australia has no more strategic warning time left. 2025 might be the equivalent of 1935. Xi and his CCP hardliners will escalate grey zone pressure on Australia in the late 2020s and early 2030s because that is all the time they have left before Australia might have a handful of submarines operating in the 2030s.

Sending traditional deterrent messages that Australia will have the military power to deter CCP ambitions to control its imagined Asia Pacific sphere of influence is fair enough. However, it holds back Australia's most significant strategic card, trade. The book's strategy moves the emphasis from military messaging to messaging that trade is more important than seeking control.

We welcome efforts to marshal the political will to spend hundreds of billions of dollars on building submarines, state-of-the-art surface vessels, aircraft, and missiles. AUKUS should be the bedrock of Australia's defence and has many benefits beyond promised submarines. However, we insist that the CCP grey zone campaign is a clear and present danger to Australia's interests now. The Australian government should hedge its longer-term sensible investment in AUKUS with an investment in grey zone capabilities that can deliver deterrence and de-escalation in two years against a grey zone campaign that has begun and could escalate rapidly in the next five years.

Conclusion

In 2024, unless Foreign Affairs, Defence, and Home Affairs have a secret collaborative strategy and are developing grey zone capabilities without informing the public, the Australian government is only in the early stages of coming to terms with the grey zone. The political priorities are leasing and building nuclear submarines, more surface vessels and longer-range missiles. The armed services (Navy, Army, Air Force) have no incentives to collaborate with other government departments or agencies in the grey zone.

Australia's three armed services will remain on the sidelines while the CCP escalates its grey zone campaign. There will be no act of war, so there will be no ADF mobilisation or employment of Australia's armed services. The Navy, Army and Air Force have not declared a priority to conduct undercover counter-hybrid, counterterrorist, counter-cyber, counter-information, counter-political and counter-WMD proliferation operations. One Army analyst opines that the Army in the grey zone would be 'punching at air' ... 'like a punch-drunk prize-fighter: swinging at shadows, wasting energy, resources and reputation'. He is right; the Army is not suited to landing a punch in the grey zone. The Navy and Air Force can patrol and control the sea and the sky around Australia, but they will not control or counter a grey zone campaign. Defence is the wrong hammer for the grey zone nail. In Part 3, we will argue for a new Response Force hammer.

Will she be right, mate?

General Douglas McArthur, the architect of the American victory in the Pacific War, wrote, 'Preparedness is the key to success and victory. ... The history of failure in war can almost always be summed up in two words. 'Too late'. Too late in comprehending the deadly enemy. Too late in realising the mortal danger. Too late in preparedness. Too late in uniting all possible forces for resistance.'

The good news is that there is still sufficient time to de-escalate in a way that complements and strengthens Australia's longer-term submarine-based deterrent strategy, with or without the United States. The book's proposal is an 'and/and' rather than 'either/or' strategy, or, for that matter, some sort of 'third way' like the Echidna Strategy or the 32-submarine ambush. This hedge investment in de-escalation is an affordable option for countering a 'light grey' campaign that the CCP has already begun with Australia and its regional allies. Its premise is that the CCP wants to succeed strategically without resorting to conventional warfare and does not want nuclear war.

Accordingly, another option is to de-escalate the CCP's political warfare in its light-grey phases and deter the possibility of escalation to hybrid warfare. Suppose lawful and sophisticated deterrence and de-escalation fail. In that case, Australia should have a fourth armed service – a full-time and part-time Response Force – [see Chapter 5, 6 and Part 3] that depends more on astute messaging for high strategic impact than conventional maritime, land and air firepower.

Let's put the doomsday cap on again. The United States and President Trump, weakened by domestic political and social challenges, decide not to defend Taiwan and risk a Third World War. In 2025, Trump negotiates with China on spheres of influence and withdraws militarily from Northeast and Southeast Asia to maintain an 'America first' US strategic defence perimeter at Hawaii, an American equivalent of Australia's strategy of looking out over its air-sea gap.

Trump's 'America First' deal with China leaves Japan, South Korea, and other Southeast Asian nations to remain neutral while China reunites with Taiwan after coercing an agreement for peace and prosperity. American allies in the Asia Pacific region succumb to grey zone coercion and sign agreements with China. Australia's Pacific

Islands' neighbours follow. Australia stands alone with its armed forces staring at the air-sea gap in China's new sphere of influence without a strategy or capabilities for defending the homeland from an escalating grey zone campaign aimed at a Hong Kong-style agreement for peace and prosperity.

Part Two – The Strategy

Detect, Warn, Deter and De-escalate
A Fourth Armed Service
Optimising Human Capital for the Grey Zone

Chapter 4 describes a de-escalation strategy to counter the CCP grey zone campaign and restore mutually respectful trading relations with China. It is war-preventing, not war-provoking. It is about reconciliation, not retaliation. It sends commercial, not military, messages.

Chapter 5 is a history lesson about a lack of Australian strategic imagination and stifled ingenuity during the 1941–1945 Pacific War. Australia has a Pacific War military history that should not be repeated, as well as lessons to learn for countering the CCP grey zone campaign. Australia does not need to copy anyone for its grey zone defence. Australia's Pacific War coast watchers, code breakers and commandos inspire Australian ingenuity and excellence in the grey zone.

Chapter 6 is a big idea for transforming Australia's ailing voluntary armed part-time military service into revitalised voluntary armed and unarmed part-time national service. Defence has neglected the Reserves for decades, and NDS2024 does not give them a meaningful job. This chapter identifies a real job for Australians who want to keep Australia safe. It outlines how Australia can marshal its best and brightest people for national resilience and grey zone defence. The ADF Reserves must transition to the Response Force Reserve and move from Defence to Home Affairs to be housed in the former Reserves' infrastructure.

CHAPTER 4

Detect, Warn, Deter and De-escalate

Introduction

Let's make three assertions to begin. First, a lack of imagination and ingenuity prevents Australia from developing a complementary sovereign defence strategy, not cost. Second, a Let's Trade, not Argue strategy – detect, stare, warn, deter and de-escalate – fits the grey zone and prevents rather than provokes armed conflict. The message is, 'Let's trade, not argue' rather than 'We are an American ally buying nuclear submarines and longer-range missiles'. Third, suppose the de-escalation strategy does not prevail over the CCP's determination to make Australia a 21st-century tribute state. In that case, the capabilities created to de-escalate will wreck hybrid warfare plans and invasion preparations.

Let's set the scene with an Australian and then an American history lesson.

An Australian history lesson

Emeritus Professor David Horner, Australia's foremost Second World War and post-Second World War military historian – our military history supremo – writes a dismal story of Australia's defence strategy development before Japan thrust south in 1941–42 and during the Pacific War. Australia squandered warning time in the 1930s and, lacking strategic imagination, adopted British and American strategies before and during the Pacific War. Some may conclude, 'All's well that ends well'. Think again.

To repeat the cycle of squandering warning time and depending on the imagination of allies rather than defending Australia using Australian strategic imagination insults Australian nationhood. Indeed, Australia's most potent weapon for defence against an aggressive regional power like China is its human capital. Australia may not have the numbers or resources to deploy a massive navy, army and air force. Still, Australians are smart enough to defend themselves in the grey zone and beyond innovatively if the nation faces an enormous navy, army and air force. Australia should engage diplomatically and apply pressure astutely well before the CCP mobilises its armed forces for invasion.

Horner's history of Australian strategic thinking before and during the Pacific War, or lack thereof, teaches several lessons. First, allowing the three services – the Navy, Army, and Air Force – or Defence bureaucrats, including former bureaucrats – to monopolise strategic thinking is risky. The services have a conflict of interest, and Defence bureaucrats fight budgetary wars, not real ones.

The three services achieved the worst of all worlds in preparation for the fight against Japan. The Army identified Japan as Australia's primary threat in 1924. Its strategy was for the Australian taxpayer to fund a large Army to defend against a Japanese invasion – a 'stay at home, muscle up, dig in and wait' strategy – known at the time as 'the counter-invasion strategy'. The Army and no one else had the imagination to project Australian land forces into Southeast Asia and the Pacific Islands. The Navy was mindful of the Japanese threat but argued for a larger Australian Navy that would join the British Navy to defeat a Japanese invasion fleet at sea well before it threatened the Australian homeland – a 'muscle up with the British Navy' strategy that involved projecting sea power from Singapore – the Singapore Strategy.

The Army and Navy debated whether the British would send a fleet to oppose the Japanese Navy if Britain were fighting a war in Europe. The Australian Air Force wanted more aircraft to fly with the British Air Force, arguing that air power had superseded sea power because aircraft could now sink ships. The Army questioned whether the British would deploy aircraft to Southeast Asia if Britain fought a war in Europe. The three services would not and could not compromise and collaborate 'jointly' for a sovereign Australian strategy against Japan because none wanted to risk their Defence budget allocation. Beating the other services for taxpayer dollars seemed more critical than collaborating to defeat Japan. Yes, we are guilty of using hindsight and are showing some disrespect. Read David Horner's books to discover these inconvenient truths. (Google David Horner books)

Sadly, the individual most invested and capable of curbing service influence over strategic thinking and the defence budget and guiding the government's preparations for war with Japan, Fredrick Shedden, Australia's star Defence bureaucrat, favoured the Singapore Strategy. Horner writes, 'He was a skilful bureaucrat, unafraid to challenge the military chiefs and usually working behind the scenes.' He fiercely opposed the Army's view that if Britain fought in Europe, its government would have insufficient will and weight to redeploy naval vessels or aircraft to defend Australia against Japan. Shedden argued that the Army should remain small as it would only be required to protect the Australian homeland against raids, not invasion. Rather than getting Australia ready to fight Japan, Shedden assumed Britain would not let Australia down and refereed the Defence budget fight between the services. When history proved him wrong, he argued later that the Americans would always intervene to protect Australia from the Japanese if Britain were unavailable – another use of hindsight.

There is no point in spending more words criticising our 20th-century ancestors with 21st-century hindsight. Still, the same conflicts of interest are at play today, and the same budgetary battles impede collaboration. History is repeating itself, and we cannot do anything about it. However, we do not need to repeat the lack of strategic vision, imagination, and ingenuity. Japan did not have a grey zone strategy in the 1930s, but the CCP had one in the 2020s. That's the battlespace where this book proposes an innovative de-escalation strategy and a Powerful Owl Response Force to back it up.

An American history lesson

The most prominent contemporary example of a liberal democracy hardening its security arrangements and chasing adversaries in the grey zone after a surprise attack is the United States. The unprecedented 9/11 attacks stunned the US Federal Bureau of Investigation (FBI), Central Intelligence Agency (CIA), Federal Aviation Agency, and Department of Defense. The US Government and its agencies had not expected this type of attack. To 'harden' for the future, the 9/11

Commission Report recommended that 'long-term success [against terrorism] demands the use of all the elements of national power: intelligence, diplomacy, law enforcement, covert action, economic policy, foreign aid, public diplomacy, and homeland defence.' This is the right approach for the grey zone.

While not declaring a strategy, the report acknowledged the requirement for the coordinated use of 'all elements of national power'. The US homeland had become a battlespace, and America's jihadist adversaries were in the Middle East. The United States needed hardened, agile and well-coordinated responses – upstream in the Middle East and downstream at home. Though not declared this way, the United States enhanced its detection, deterrence, and response capabilities to prosecute the War on Terror in the grey zone. The government combined 22 federal departments and agencies into a unified, integrated Department of Homeland Security (DHS) in 2002. In 2005, the government simplified the DHS to include border and transportation security, emergency preparedness and response, information analysis, and infrastructure protection. It took Australia another 12 years to consolidate homeland security.

The United States sharpened its instruments for both creating deterrence and striking at terrorist networks at home and abroad. The government enhanced the capabilities of the FBI, the CIA, and special forces to ensure that there were 'ultimate sanctions' for terrorism at home and abroad. In effect, the Americans created sufficient deterrence and de-escalation capabilities at home – downstream – with the DHS and the FBI and abroad – upstream – with the CIA and special forces to counter terrorist jihadist networks in the grey zone. We are not arguing for or against US decisions to intervene militarily in the Middle East. Our point is that the US and its allies went after terrorist networks 'upstream' and 'downstream'.

Australia did not endure a galvanising attack like 9/11 or horrific terrorist bombings like those in Britain and Europe in the 2010s. Australian governments have responded reflexively and incrementally to nano-9/11s, such as the 2014 Lindt Café Siege and the 2017 Brighton Motel Siege and other 'lone wolf' incidents in Melbourne and Sydney and the deaths of Australians in overseas terror attacks in Bali, London and Bagdad. Australia's legislative and institutional responses to people smuggling and border security also prompted the consolidation of security agencies into Home Affairs. Australian special forces joined their American and British counterparts in the Middle East for the War on Terror in 2001 and were still serving there in 2024.

The book does not advocate following American precedents. Australia does not need national security instruments like the FBI and the CIA or to emulate them as models for how Australia should defend itself in the grey zone. Australia can enhance existing departmental and agency structures to align authority, responsibility, accountability, and resources astutely and strengthen law enforcement, special operations, intelligence, and national security machinery. We can and should 'do our own thing'.

Australia is not countering terrorism in the 2020s with the same intensity as in the 2000s. The primary threat comes from Xi Jinping and the CCP's aspirations to change the world order to suit themselves. Still, the enhancement of existing organisations and processes will not be enough. Australia needs a whole-of-government de-escalation strategy for a bullied middle power to stand up to a bullying superpower but avoid war. It is a strategy for returning China and Australian relations to mutual respect and free and fair trading arrangements. The significant structural change is that this strategy requires a fourth armed service and transforms Australian part-time military service from an Army Reserve preparing to defend against invasion to part-time national service, bolstering national resilience against natural and grey zone disasters (Chapters 5 and 6).

De-escalation

The book's grey zone strategy is about de-escalation – a fair and firm assertion of Australia's sovereignty (independence). The concept is not new. It is firm behaviour to stop an escalation of bullying behaviour. It is about proportional and lawful responses to coercion. It would not be needed if the CCP wanted to trade and not interfere and control. The aim is to thwart escalation from non-violent political pressure to violent, disruptive and destructive intimidation and hybrid warfare.

More particularly, it is the employment of detection, deterrent, information and de-escalation organisations to anticipate and counter the three phases of the CCP grey zone campaign at home and abroad. Different departments and agencies engage during each phase to detect, warn, and deter. Responses from a new response force become more sophisticated and lethal if the CCP ignores the invitation to trade and escalates violence, disruption, and destruction.

The book's strategy comprehends Australia as a middle power and builds on Australian values and traditions. At home – 'downstream' – it builds on a foundation of mature counter-terrorism arrangements. Abroad, – 'upstream' – builds on an Australian tradition of special operations that began but did not achieve full potential in the Pacific War. It is mindful and respectful of Australia's trading relationship with China, the mutual obligations of the US alliance, and relationships with regional neighbours. The CCP has been rude and nasty, and they should know that this behaviour towards Australia will have uncomfortable consequences in the future.

The de-escalation strategy aims to cause CCP hardliners to have second thoughts and encourage CCP moderates and China's corporations to argue for ending aggressive statecraft and returning to fair, respectful and equal relations with Australia based on mutually beneficial trade. It detects and assesses changes in the CCP's attitude and intentions for Australia. It develops and delivers astute warnings, and if warnings and the knowledge of Australia's agile countermeasures do not deter coercive escalation, the strategy is to de-escalate with uncomfortable 'jabs' and 'stings' to prompt a return to negotiation and reconciliation. If 'jabs' and 'stings' do not stop escalation, Australia has additional countermeasures to forestall hybrid warfare.

This proposal is not a replacement strategy. Australia's defence strategy is politically and diplomatically sensibly based on advantageous geography, traditional priorities (homeland, region, world), and a strong ADF. The proposed de-escalation strategy counters grey zone campaigns, complements Australia's military deterrence and alliance arrangements, and will support traditional military campaigns if war is declared.

Imagining Australia's defence strategy as a chair, the advent of political and hybrid warfare in the grey zone leaves it with only three legs – maritime, land and air power. The fourth leg of Australia's defence strategy chair should be Response Power. It incorporates counter-cyber and counter-information warfare capabilities and optimises existing civil-military security machinery for counterterrorism and countering foreign interference. It develops detection, deterrence and de-escalation capabilities to counter grey zone operations anywhere and anytime. It has Powerful Owl attributes of intelligence, astute detection, ominous warning messages, stealth and long-range surprise with amplified informational impact in the first instance to send more potent messages or – as a last resort – 10 times amplified disruption of hybrid warfare and invasion preparations.

This strategy responds to the current 'light grey' CCP campaign in Australia and its neighbourhood in the first instance. In exchange for removing trading embargoes and softening diplomatic language, the CCP expects Australia to join it in condemning Donald Trump's threats of trade tariffs on China to coerce better behaviour. Hopefully, moderate CCP factions and China's corporate leaders will recognise, understand and grow to respect Australia's measured and polite refusal to distance itself from the American alliance. CCP hardliners must learn to respect Australia, an important trading partner, and concede and accept Australia's separate existence as a liberal democracy and entitlements to its sovereign rights and alliance choices.

The strategy dampens incentives or temptations for the CCP to escalate to darker-grey hybrid warfare. While understanding that Australia must communicate carefully and diplomatically with one of its major trading partners, cyberattacks and espionage, political interference and intimidation of Australian citizens warrant more than firm condemnation. Australia must warn the CCP that bullying Australia has hurtful consequences. The messaging tone is disappointment, not anger, and intends to dissuade those seeking inappropriate influence, i.e. Australia seeks reconciliation, not retaliation. Ultimately, there is no point in being disappointed, asking for meetings, making conciliatory speeches during diplomatic visits and issuing warnings without the capability to back them up with deterrent actions.

The ultimate deterrence and de-escalation dimensions of this strategy are simple. Australia must have a fourth armed service, the Response Force, to conduct extraordinary operations anywhere and anytime to de-escalate 'dark grey' threats. Intense political, diplomatic, and informational engagement that seeks reconciliation will always accompany Response Force de-escalations. It disappoints rather than provokes – more on Response Force's extraordinary operations in each phase of a grey zone campaign in Part 3.

Intentions

The strategy defends against threats to national diplomatic, economic and political interests in whatever form they take at home and abroad. Its Response Power dimension provides the ways and means for acquiring intelligence, enabling precise targeting, conducting preliminary non-violent disruptive operations and harder de-escalation actions if deterrence fails. It maintains Australia's non-nuclear policies for military capabilities while at the same time defending Australia against nuclear and other WMD threats by responding to them before they reach dangerous thresholds or imminent use.

This strategy's front line is 'anywhere', and the readiness level is 'anytime'. The aim is to engage threats 'upstream' overseas rather than risk terrorist-like surprises 'downstream' at home accompanied by disruptive cyber-attacks. It offsets reliance on the ADF or dependence on allies to respond. It gives Australia a sovereign option to shape, understand, influence, and respond to threats at their different stages of development.

This strategy complements collaboration with allies. It enables Australia to join partners in operations in the grey zone to protect the rules-based global order and de-escalate potential conventional or nuclear war threats. It gives authority and direction for developing and employing the Response Force to counter grey zone campaigns before they escalate.

Force is the last resort. Australian legislation, the Laws of War, the Geneva Conventions and human rights conventions bound this strategy. Australia is a liberal democracy founded on respect for human life and the rule of law. Lethal and destructive force is not the first or the only option. The emphasis is on messaging, prevention, protection and deterrence. It is a graduated response process that 'paces' the shades of grey lawfully. Intense diplomatic and political engagement to find peaceful consensus accompanies each step towards the possibility of applying lethal force. While regretting the requirement to develop Response Power, it is a strategy that does not shirk from forcefulness. The intention is to detect and disrupt threats in the making. If deterrence and disruption fail, de-escalation will hurt.

This strategy does not require ethical compromises. However, Australian governments must take decisive action in response to illegal coercive intentions and activities. It is a firm response to unlawful behaviour, such as espionage, political interference and intimidating Australian citizens in their own country. It will involve lethal force if those seeking to dominate Australia escalate to violent and destructive actions.

Response power

Response power is the practical extension of the NDS2024 concepts of shape, deter and respond. The Response Force (hereafter RF) hardens law enforcement, diplomatic and grey zone capabilities for Australia's national security. It paces and responds to Phases 1, 2 and 3 escalations. It is a spectrum of responses to detect, pace and counter an escalating grey zone campaign through these phases.

It is founded on prevention rather than provocation and de-escalation rather than retaliation. Response Power integrates with Australia's diplomatic, informational, military and economic power. It intends to give firm and fair consequences for coercive behaviour against Australia's sovereignty and national interests while inviting the CCP to trade and not argue. It upholds liberal democratic values and the rule of law but is potent if negotiation and deterrence fail and forceful de-escalation is required.

Response power has law enforcement, diplomatic and extraordinary operations instruments that rely on the NIC to detect threats. Home Affairs and Foreign Affairs are the national and international information and negotiation pillars. Response force is the deterrence and de-escalation pillars.

Diagram 1

Pillars of Response Power		
Detection	Information Negotiation	Deterrence De-escalation
National Intelligence Community	DFAT Home Affairs	Response Force (full & part-time, domestic and international)
	Anywhere and Anytime	

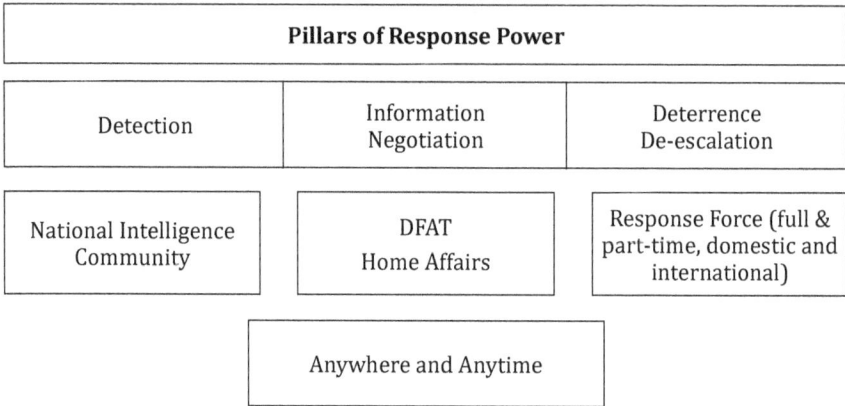

The Response Power Cycle – who and how

Response Power generates strategic pressure and opportunity. The means are a full-time and part-time RF that is politically, diplomatically, technologically and culturally astute. The RF will have a full-time and part-time workforce employing cyber and space capabilities, information technologies, extraordinary operations, and autonomous and remote-controlled weapon systems. The book does specify these capabilities for security reasons.

In the first instance, Response Power hardens Australia against foreign espionage, political interference, cyber-attacks and terrorism. It requires the extension of current security legislation to authorise multi-agency teams from several levels of government to meet complex grey zone challenges at home and abroad. A coalition of government departments and agencies, already working together for counter-terrorism, will conduct law enforcement, cyber, informational and electronic operations simultaneously or sequentially against grey zone combatants.

The de-escalation cycle involves detection, 'staring' (continuous overwatch), analysis, decision, communication/warning, and action, if required. The NIC detects and stares at home and abroad. Depending on the level of danger, the RF supports law enforcement agencies and the NIC 'staring' at existing and emerging threats via covert surveillance and reconnaissance. At home, a Home Affairs-led

ASIO/AFP/Australian Border Force (ABF) partnership and the RF's domestic full-time and part-time components assigned from PM&C do the detection and staring. Foreign Affairs, ASIS, and the RF overseas component detect and stare (surveillance operations) in the near region and internationally. (More on what the RF will do in Chapter 8)

Another important 'who' for detecting, staring and reporting are members of the 1.2 million Australians of Chinese heritage. The CCP, especially the United Front, has been targeting them and infiltrating their media and cultural organisations to enlist members to support China's aspirations to change the world order. China has commercial and family links to this community, and the community has commercial and family links to China. These Australians understand and have experienced the CCP grey zone campaign more painfully than any other community in Australia. The NIC and the RF must recruit and train selected volunteer members to counter grey zone escalations within the community and assist with Australia's de-escalation cycle at home and overseas; Australia and China benefit from their participation.

While the RF and its partners stare and report, the Australian government decides how and when to communicate warnings that there will be consequences for planning threatening behaviour or conducting illegal grey zone operations. Home Affairs communicates and acts domestically, and Foreign Affairs does the same internationally. Sophisticated warnings to achieve information dominance are crucial for achieving deterrence. This dominance is the ability to collect, manage, and exploit accurate information more rapidly than anyone else. The government shapes domestic and international public opinion and perceptions to counter CCP bullying by outpacing the CCP's information operations inside Australia and overseas.

Warnings aim to deter light-to-dark-grey escalation, but if the CCP ignores warnings and persists in sponsoring and perpetrating illegal behaviour, there needs to be real and ready deterrent action. The CCP may stop bullying behaviour if it knows about uncomfortable consequences, but if knowing about them is not enough, then the CCP needs to 'feel' consequences. Deterrence begins with well-crafted political, societal and diplomatic communications and negotiations to influence CCP attitudes and create dilemmas, wicked problems and

Catch-22s. But if coercive behaviour escalates, deterrence must be real, and the CCP must 'feel' it.

A Home Affairs-AFP-ABF Response Force deters and de-escalates for Phase 1 in the homeland to counter a light grey zone campaign operating illegally below the threshold of coercion. This combined effort is 'business as usual' law enforcement applied under existing legislation. Collaboration changes when darker grey unlawful Phase 2 tactics are detected. Cyber-attacks become more disruptive. Trade embargos increase. Diplomatic and political messaging becomes more aggressive, and espionage, recruitment of individuals for violent action and political interference intensifies. The RF brings additional detection surveillance technology to bear, and de-escalation rehearsals begin. Now called out and engaged, the RF component that has been selected, trained and rehearsed for extraordinary operations de-escalates Phase 2. The RF raids facilities and grey zone combatants end up in gaol. If the CCP continues to escalate, the RF is poised after Phase 2 operations to counter a Phase 3 dark grey zone campaign – hybrid warfare – if it escalates to violence in Australia's homeland.

This graduated and carefully calibrated response cycle shows how the RF paces the three phases of a grey zone campaign. It is a de-escalation cycle, not a retaliation cycle, but it does intensify and 'over-matches' escalation by being pre-emptive and explicit in its messaging. The CCP must conclude that the consequences of escalation are too uncomfortable and that they are being detected – sprung – every step of the way. i.e. named and shamed with an accompanying invitation to trade and not argue.

Concurrently, the RF Foreign Affairs-ASIS overseas component detects and stares at grey zone threats and counters a light grey Phase 1 campaign. It will pace threats as they transition to darker grey Phase 2 and 3 tactics with warnings, stings, and jabs designed to de-escalate and seek reconciliation and restoration of trading relations.

The RF is 'always on' with hurtful, non-violent and violent capabilities. The 'stings' and 'jabs' to get attention for negotiation are intelligence-led and carefully calibrated extraordinary operations (see Chapter 8). They aim to surprise, shock and de-escalate. Importantly, action is always accompanied by intense political and diplomatic engagement seeking reconciliation. De-escalation is about constraint – not containment – and diplomatic and political engagement – not adversarial exclusion or isolation. The message is 'Let's trade, not argue'.

Stinging and jabbing bolsters deterrence. Initial stings may be the public exposure of a plot. If the CCP ignores public exposure and international criticism, the arrest and detention of operatives or a raid on a facility to arrest and detain grey zone combatants accompanied by publicity to optimise the exposure of illegality are options. The CCP may have second thoughts about escalation if well-publicised red kangaroo stencils appear inconveniently and alarmingly on sensitive Chinese facilities or during CCP public events in China and elsewhere. As a last resort, a sting may cause loss of life and destruction. Still, it must always come after the CCP is about to take, or has taken, Australian lives and destroyed Australian property. Even in these circumstances, there should be a fair warning and an invitation to negotiate unless surprise and safety are paramount.

Ideas for regional de-escalation

Australians should not stand alone in the grey zone but persuade allies to join them for de-escalation. Australian strategist Ross Babbage's prescriptions of 'partnership and leverage', and former Foreign Affairs Secretary Peter Varghese's policy guide rails of 'engage and constrain' for implementing a regional de-escalation strategy apply. Babbage states,

> At its core, the strategy of partnership and leverage [with regional allies] would also substantially enhance deterrence and defensive capabilities against any attempt to coerce or attack ... [It] offers powerful leverage to deter regional bullying and force any aggressor to cease operations and quickly come to reasonable terms.

Peter Varghese opines,

> [The question is] how to deal with a China that is not interested in reaching mutually agreed rules. What do we do then? That is why I

think we cannot have an honest narrative that does not also canvass the need to constrain China's unacceptable behaviour through some *new balancing arrangement* [author's emphasis], which sits side by side with [continuing] engagement. To think we can leave it with striving for mutual agreement on rules leaves us open to being seen as naive.

The Singaporean statesman Lee Kuan Yew's 'poisonous shrimp' strategy in the mid-20th Century to deter larger Asian powers from interfering in Singaporean affairs and seeking to make Singapore a tribute state is instructive for Australia. Ironically, he drew on a Chinese proverb when he said at a conference in 1966 that:

In a world where the big fish eat small fish, and small fish eat shrimps, Singapore must become a poisonous shrimp. ... There are various types of shrimps. ... Species in nature develop defence mechanisms. Some shrimps are poisonous: they sting. If you eat them, you will get digestive upsets.

An attendee at the conference asked what he meant in the context of Singapore's dependence on trade and service delivery with its 'big fish' neighbours, China, Malaysia and Indonesia. He replied, '... our separate existence having been accepted and conceded, we then deal with them on equal and fair terms'. A bullying CCP will not accept Australia's 'separate existence' until there are painful consequences for escalating in the grey zone.

A de-escalation strategy aims for Australia and its neighbours to deal with China equally and fairly if China respects their sovereignty. Such preparations would be unnecessary if China's intentions for Australia and its neighbourhood were benign. If the CCP persists with its grey zone campaign, Australia and its regional allies should develop defensive capabilities to become 'poisonous shrimp'.

The book's de-escalation strategy combines Babbage, Varghese and Lee Kuan Yew's advice with Australian diplomatic efforts to strengthen political, economic, defence and border security relationships with regional neighbours. The focus would be enhancing regional capabilities to resist the CCP's illegal interference and coercion. The imaginative development and employment of regional RF trained by and partnered with the Australian RF will create a potent deterrent for the CCP's light grey Phase 1 and 2 campaigns. If a Phase 2 escalation occurs, Australia's RF can assist. Figuratively speaking, the Chinese big fish swimming in the Asia-Pacific region will find it uncomfortable to digest schools of unpalatable stinging shrimps.

The book's strategy for regional engagement with response forces is not new. The idea of responding anywhere and anytime in the company of allies originated in Australia's initial National Counter-Terrorism Plan in 2001 and is present in the current plan:

> The AIC [Australian Intelligence Community or NIC] and the AFP maintain overseas liaison channels to gather intelligence and pursue investigations. ASIO and the AFP also maintain a 24-hour monitoring and alert unit. Relevant Commonwealth agencies provide the interface with overseas security, intelligence and police agencies as part of international counter-terrorism efforts. Defence contributes to the prevention of terrorism through its overseas operations by disrupting a primary source of, and inspiration for, terrorist activities worldwide.

This targeted and deliberate enabling plan with regional counter-terrorism forces is a prototype for collaborating with grey zone response forces. It commits Australia to liaise and share burdens with countries countering CCP grey zone campaigns. The Australian government has authorised its special forces to target terrorist groups in their foreign sanctuaries in conjunction with allies and local civil and military authorities. This authorisation countered terrorists 'up-stream' rather than 'down-stream' in Australia, where their destructive plans might be too late to stop. How Americans would have rejoiced had their special forces intercepted and neutralised those who perpetrated 9/11 'upstream'.

Australia's de-escalation strategy is exportable. Australia should continue its 'upstream' strategy for countering terrorism anywhere and anytime by assisting regional neighbours to harden up against grey zone campaigns. This strategy follows traditional strategic prescriptions of affording Australia's neighbourhood a high priority for defence after defending the homeland first. The logic is that countering a grey zone campaign in the near region protects Australia, and what defends Australia in the grey zone should also protect its neighbourhood.

Ethics

Solid moral and ethical messages accompany the Response Power Cycle. It is incumbent on a liberal democracy like Australia to declare its intentions to de-escalate coercive and illegal threats against its national interests with force. It does so for all other threats. Australia will be employing its RF lawfully in compliance with international laws on armed conflict and human rights. The Attorney General must scrutinise all actions under the Human Rights (Parliamentary Scrutiny) Act 2011. The Laws of Armed Conflict incorporated into the 1949 Geneva Conventions, especially Common Article 3, which covers situations of non-international armed conflicts, must apply. Also applicable is the International Covenant on Civil and Political Rights and the Convention Against Torture and Other Cruel, Inhuman or Degrading Treatment or Punishment.

The book's strategy is carefully calibrated to be below the threshold of war. It is like a cocked fist. Its stinging jabs are a last resort in the face of the threat of force or other warlike actions. It responds to coercive acts that will cause harm to Australian citizens directly or indirectly through the destruction or disruption of essential services, supply chains and economic activity. Before and after every jab, there is an invitation to talk. It admires the boxer Muhammad Ali's boast, 'Float like a butterfly, sting like a bee. The hands can't hit what the eyes can't see.'

Conclusion

This chapter describes a de-escalation strategy and Response Power Cycle that are war preventers rather than war provokers. It calls for including Response Power with maritime, land and air power to engage and de-escalate the CCP grey zone campaign. For Phases 1 and 2, the strategy requires the Australian government to build on counter-terrorism legislation and arrangements to deter terrorist-like dark grey escalations at home and abroad. ASIO and the AFP must be returned to the Home Affairs portfolio. The only significant change is establishing an armed component of the RF. (See Chapters 7 and 8)

Australia's Phase 3 grey zone defence responds to an escalation to dark grey hybrid warfare. The Phase 1 and 2 national (Home Affairs/ ASIO/AFP/ABF) RF and international (Foreign Affairs/ASIS) RF need to establish capabilities as a fourth armed service that does what the Navy, Army and Air Force are not suited or willing to do. This Phase 3 capability is a secretly prepared and rehearsed RF component.

The following two chapters explain, through another Australian history lesson and a big idea, why a fourth armed service – a Response Force – is necessary and must have its roots in part-time voluntary national service that marshals Australia's human capital for its grey zone defence.

CHAPTER 5

The Fourth Armed Service

So far, we have shown that there is a grey zone threat and what might happen if it is not de-escalated. Australia's current military strategy does not face the danger squarely. Still, a solid intelligence and law enforcement foundation exists to build a grey zone defence in the Australian homeland. The last chapter described Australia's national security strategy as a three-legged chair – maritime, land and air power – that requires a fourth leg – response power. It proposed a detection, warning, deter, and act strategy for meeting the threat 'upstream' overseas and 'downstream' in the homeland. This chapter proposes a fourth armed service – Response Force – to be that fourth leg, joining the Navy, Army and Air Force legs to defend Australia.

Before determining where an RF would fit in, how it would be employed, and what it would need in Part 3, the book sets the scene for this fourth armed service with a history lesson and a big idea. The history lesson in this chapter is that Australia had a prototype fourth armed service during the Pacific War that did not achieve its full potential. The big idea in the following chapter is that Australia's people – the quality rather than the quantity of the nation's human capital – should be mobilised with a new vision of voluntary part-time national service for national resilience against natural disasters, civil emergencies and grey zone catastrophies.

A history lesson about intelligence and special operations
In the 1930s, Australia depended on, and in the 2020s, it still depended on the three services – the Navy, Army, and Air Force – for its defence. Each service trains to fight and defeat an enemy's navy, army and air force. Their capabilities constitute Australia's maritime, land and air power. In the 1930s, Australia depended on, and in the 2020s, Australia still relied on a powerful ally to defend Australia against attack. In the 1930s, Australia put its strategic faith and hope in Britain. It adopted

the Singapore Strategy, in which Britain would deploy sufficient maritime, land, and air power from Europe to defend Australia against attack. In the 2020s, Australia put its strategic faith and hope in the United States under the terms of the ANZUS Treaty and the goodwill of presidents and ruling parties to deploy sufficient maritime, land, and air power to defend Australia against attack.

While Australia's three services developed as allied dependencies, Australian intelligence services and special operations capabilities developed independently in response to threats to Australia and its national interests. We will define contemporary covert and clandestine intelligence and special operations, which we will rename 'extraordinary' later in Chapter 8. For now, let's discuss the origins and story of Australia's development of its intelligence and special operations capabilities leading up to and during the Pacific War in the 20th Century to inform the development and employment of a Response Force in the 21st Century.

Australia began intelligence gathering and special operations early. Australia's first-ever clandestine intelligence operation was launched against France and Britain in the New Hebrides [Vanuatu] a few months after Federation in 1901. This operation demonstrated that Australia was smart enough to protect its national interests – independent of allies and against any rival nation's encroachments in its near region. Atlee Hunt, Prime Minister Edmund Barton's private Secretary and, concurrently, Secretary of the Department of External Affairs [now Foreign Affairs], was Australia's first spymaster. Even after a change of government in 1910, he continued as the Secretary of the Home Affairs Department with border protection responsibilities. He established an extensive national and international intelligence gathering and reporting system incorporating State governments, police forces, those managing harbours, customs officials, and State trade officials deployed overseas. Australia had eyes and ears in its neighbourhood and beyond.

While Hunt developed the first sovereign intelligence network in the decade after Australia's Federation, Australia had no separate organisation dealing with foreign espionage, subversion or sabotage in the homeland; in its absence, the Department of Defence took on the responsibility. French colonies in the Pacific Islands remained the focus of Australia's covert operations. Major William Bridges, later a Major General in the First World War, became Australia's first undercover spy posing as a Dalgety commercial agent in French New Caledonia. After Japan's victory over the Russian Navy in 1905, Australia developed naval intelligence capabilities that leveraged the Royal Navy's worldwide intelligence network. Australia's maritime intelligence would go on to achieve the first intelligence coup of the First World War – the seizure of Germany's naval codes from a base in the southwest Pacific Islands in 1914. The fledgling Australian government was intelligent, adaptive, and independent while still a loyal British ally.

The First World War prompted modest increases in Australian intelligence services for opposing Germany and keeping the Australian homeland safe from German espionage, subversion and sabotage. The Australian Army's Directorate of Intelligence assumed international and domestic intelligence-gathering responsibilities. After Britain established a Counter-Espionage Bureau [later called MI5 and MI6], the Australian Governor General's Secretary, Major George Steward, custodian of the Governor General's secret ciphers used to communicate with London, headed Australia's equivalent that was renamed the Australian Special Intelligence Bureau (SIB) in 1916. In 1917, the Federal Government established the Commonwealth Police (later called the AFP) to detect subversion by monitoring organisations such as the Industrial Workers of the World and later the nascent Communist Party of Australia (CPA).

There were several lessons from the first two decades of Australian nationhood. Sovereign intelligence services were necessary because Australian and allied interests were not always aligned. All nations pursue sovereign national interests first. Australia did not coordinate intelligence capabilities efficiently or effectively; the organisations established to meet threats to Australian security had different lines of responsibility and jurisdictional divides that limited collaboration, and there was no cohesive border control. Intelligence services should be led at the highest level, and there are roles for both civilian and military departments and agencies. There were impressive results when human intelligence [Australian spies and those paid to spy for Australia] and signals intelligence [the Australian Government listening to other people's telecommunications] combined. Hold that thought.

While the need for domestic Australian intelligence services was self-evident and affordable, international intelligence services were expensive and required political, cultural and linguistic smarts. Australia tended to depend on the British intelligence services for global operations. Australian governments and their armed forces struggled to allocate sufficient resources for language training. Even with the benefit of hindsight, the creation of Japanese language capabilities during the First World War and subsequently when Japanese nationalism was on the march was haphazard. By 1921, a civilian Pacific Branch conducted brief language courses for military officers, diplomats and spies. The government disbanded this branch and ceased language training, leaving the armed services as the nation's only active international intelligence collectors. Australia was the recipient of British intelligence, not the creator of its intelligence. The lesson is that Australia should never allow the equivalent relationship to exist in the 21st Century.

During the interwar years 1919–1939, Australian governments reduced defence expenditure and focused their domestic intelligence organisations on 'staring' at Australian Communists. The division of labour between the civilian Commonwealth Investigation Branch, established in the early 1920s, and military intelligence was never satisfactorily resolved.

The Pacific War

By 1939, when Australia joined Britain in declaring war on Germany, Australia did not have an effective, all-encompassing counter-espionage or security service. The domestic priorities of wartime, namely, the internment of aliens [persons of German and Japanese heritage], censorship, and protection against subversion, espionage and sabotage remained. Technology facilitated radio interception, making signals intelligence and acquiring foreign language skills crucial. Fortunately, and importantly, Australia, an island nation dependent on seaborne trade, did have an impressive network of volunteer coast watchers. The Australian Commonwealth Naval Board established a network of 700 volunteer coast watchers around the Australian coast and in the islands of the near region to the north and northeast of the continent. … 'one of the most effective and reliable human intelligence systems operated during the Pacific War'. Hold that thought.

The Pacific War was a massive psychological shock for Australia. In high anxiety, the Australian people realised that their governments had squandered the 1930s warning time and were unprepared for war in their near region. Until the attacks on Pearl Harbour on 6 December 1941 and the rapid advances of the Japanese armed forces south, Australia had not faced an Asian enemy capable of conventional warfare. Australia hadn't developed a deterrent strategy or established strong enough co-dependencies in trade to deter a significant trading partner from conquest and possible invasion. Now faced with that threat and the possibility of homeland espionage, subversion and sabotage, Australia was unprepared, partly because of dependence on the British Singapore Strategy but mostly because Australian governments did not think it through in the 1930s. There was plenty of information about Japanese aspirations in Southeast Asia. The Australian Government and media debated the risks of Britain not protecting Australia against the Japanese nationalists but did not convert that information and debate into developing a sovereign strategy to complement the Singapore Strategy.

Australia's vulnerability increased after its allies decided to 'Beat Hitler First' in 1942. Britain had no spare maritime, land or air power to defend Australia. Prime Minister John Curtin quickly switched Australia's alliance allegiances and called for American military power to thwart the Japanese armed forces. Thankfully, the Japanese had blundered with the Pearl Harbour attack and galvanised the full might of the United States into action against them. The United States and Australia had a shared enemy. Still, Australia relied on American imagination and military weight to defeat Japan. Australia mobilised for its defence and looked to repatriate its maritime and land forces, pilots, and aircrew serving in Europe and the Middle East. But what of its sovereign responsibilities for intelligence and special operations?

Japan had not conducted special operations against Australia in the interwar years, except for covert surveying of the Australian northern shoreline. However, with war declared, theoretically, the Australian homeland became a target. The Japanese homeland, Japan's armed forces and national interests in Southeast Asia and elsewhere also became targets. Australia's homeland intelligence services were a shemozzle because civil and military responsibilities were unclear, and agencies were separate 'silos'. The CIB, Military Intelligence and co-opted State and Territory police forces, and the new Department of Information all competed and squabbled to address national security priorities of the internment of aliens, censorship, and protection against subversion, espionage and sabotage. Fortunately, Japan had no sophisticated special operations capabilities or technology for a grey zone campaign to exploit Australia's unpreparedness and lack of coordination. Hold that thought.

A new Commonwealth Security Service (CSS) directed by the Attorney General's Department and operated by the Army with deputy directors in each State tried to consolidate intelligence collection and dissemination in the homeland. There was a deep ill feeling between the CSS and the armed services' security organisations. Fortunately, the Americans were ready to 'bang heads together' and direct the efforts of Australian intelligence agencies who were not cooperating. Under overall American command, Australian military intelligence folded into American-controlled organisations, leaving civilian intelligence agencies to focus on essentially non-existent homeland threats from Japan and watching communists. The civilian CSS achieved uneven results in counterespionage, counter-sabotage and counter-subversion. The lesson for Australian intelligence services was that centralised control from the top works, but authority, responsibility, accountability and resources between civilian and military organisations must be precise.

The most significant organisation in terms of winning the war against Japan was the Allied Intelligence Bureau, which broke Japanese codes and intercepted Japanese military radio traffic. Australia had brilliant people who could intercept, decrypt, and analyse Japanese telecommunications. Hold that thought. Still, Australia had no special operations capabilities to hurt the Japanese war effort after Japan struck the first blow at Pearl Harbour.

Fourth Fighting Force 1942–1945

It was not until 1942, months after 'the balloon had gone up' that Australia's intelligence services and special operations began maturing to meet the Japanese threat. Before picking up this story, it is instructive to roll back to 1940 in Britain to see how British Prime Minister Winston Churchill single-handedly jump-started British special operations. None of the Service chiefs was either interested in or made responsible for learning about and hurting the German war effort after the declaration of war in 1939 and in occupied Europe after Dunkirk in 1940. Churchill founded the Special Operations Executive [SOE], a fourth covert armed service, in July 1940 after the Dunkirk evacuation, directing the SOE to 'Set Europe ablaze' . In the language of the time, 'SOE was considered by the War Cabinet as a separate service, a service able to carry out tasks that the other services could not undertake. For example, the employment of [British] civilians, foreign nationals and native peoples [locals] for subversive activities.'

SOE was a national asset with separate government funding. It was directed by the Minister for Economic Warfare and formed 'to coordinate all action, by way of subversion and sabotage against the enemy overseas.' The British Navy and Air Force resented SOE, calling it 'Churchill's Secret Army' and 'The Bureau for Ungentlemanly Warfare'.

By the war's end, SOE had a staff of 13,000 personnel. Allied senior officers of all three services widely praised its operatives for shortening the war and saving Allied lives in Europe after the D-Day invasion. SOE terrified the Germans with special operations that not only 'blew stuff up' at ports, bridges, railway lines, communication hubs and airfields but also killed selected individuals, disrupted supply chains and communications, mobilised and armed locals and spread propaganda to unsettle the minds of German military personnel.

An Australian prototype fourth armed service, emulating Churchill's SOE, began haphazardly forming in 1940. Initially, Australia received British guidance for 'para-military activities, including raids, demolitions, and organising civil resistance and sabotage' behind enemy lines. Mission 104 was sent from Britain to Australia in December 1940 to raise and train elite independent companies that evolved into Australia's first special forces commando companies. The British War Office did not disclose its intention for Australian independent companies to reinforce British commando operations in Europe and the Middle East. Mission 104 was thrown out of Australia because of its duplicitous and condescending behaviour and the discovery of its intention for Australian commando companies to defend Britain, not Australia. Britain had been caught harvesting Australia's human capital – intelligent and talented Australians – for its use, as was the case with pilots, aircrew and intelligence staff.

Mission 104's legacy was Australia's first special forces commando companies for the Pacific War and a training base and curriculum for special operations at Wilson's Promontory on the southern coast of Victoria. In Australia, as in Britain, the three service chiefs and American senior officers resented supporting special operations, preferring conventional warfare against opposing navies, armies and air forces. Hold that thought.

Like Britain, leadership for establishing a special operations organisation in Australia came from the top. The Australian Government raised the Inter-Allied Services organisation that later became Special Operations Australia [SOA] in March 1942, four months after the Japanese attack on Pearl Harbour, under the command of General Thomas Blamey, Command-in Chief, Australian Military Forces [equivalent of the contemporary Chief of the Defence Force (CDF)].

Prime Minister Curtin kept SOA separate from the services and set aside funds for its establishment. But this organisation and its successors soon came under American control. They were accountable to American approval processes and resource managers for allocating submarines and aircraft for commando and agent deployment.

From the end of 1942, General MacArthur's General Headquarters took control of all US, Australian, and allied special operations but did not prioritise them. In April 1943, SOA, disguised as the Services Reconnaissance Department [SRD], reported to the Controller, Allied Intelligence Bureau, an Australian with an American deputy responsible for managing resources. All special operations had to receive General MacArthur's approval. He suspected British and Dutch special operations disguised post-war colonial agendas to reassert control.

The development of Australia's SRD capabilities was slow. Australia's three armed services were not interested. It took until October 1943 to establish the Fraser Island Commando School near Brisbane, a more fitting climate for Southeast Asian operations, and for trainees to rehearse raids on airfield, port, and railway infrastructure. SRD received low resource priority and depended on navy, army, and air force training units for specialist tactical and language training.

During this time, SRD began developing special operations units. Z Special Unit operated behind Japanese lines in Southeast Asia. Predominantly Australian, the Z Special Unit was a specialist reconnaissance and sabotage unit that included British, Dutch, New Zealand, Timorese and Indonesian nationals. It operated in Borneo and the islands of the former Netherlands East Indies [Indonesia and Timor Leste]. 'M' Special Unit was an SRD Allied reconnaissance team. It was the more formally structured successor to the highly successful coast watchers' network, which had played a vital role in the early stages of the Southwest Pacific campaign, culminating in the Battle for Guadalcanal in Solomon Islands.

No substantial SOE was developed for special operations in North and Southeast Asia in the first years of the Pacific War. Even after the tactical success of Australian commandos in Timor [now Timor Leste] and Z Force in Singapore Harbour in 1943, there were no innovative minds above the tactical level capable of turning tactical experience and success into strategic success. Based on David Horner's analytical histories of higher Australian command during the Pacific War and biographies of General Sir Thomas Blamey and Sir Fredrick Shedden, Australia's Secretary for the Department of Defence, there was little interest in developing sovereign Australian intelligence services and special operations for strikes against Japanese forces occupying Southeast Asia. With the benefit of hindsight, senior Australian leaders did not recognise that investing in commandos, intelligence agents, and strike teams was affordable, innovative and could be strategically effective.

It took over 18 months to raise and train commando companies and Z Force and M Force covert teams to conduct raids and enable operations among local populations in Southeast Asia. Australia's first successful raid on Singapore Harbour occurred in late September 1943, two months short of two years after the surprise Japanese attack on Pearl Harbour. 'The raiders were credited with sinking seven ships weighing about 40,000 tons total and shattering the Japanese myth of invincibility in the port of Singapore'. By any assessment of investment return, this was a low-level tactical operation that resulted in a strategic-level effect. The investment was 14 specially selected and trained commandos, one fishing boat, several canoes and a couple of dozen limpet mines. Metaphorically, this success was an extraordinary Australian Powerful Owl operation that killed prey more than ten times its body weight. Hold that thought.

This lack of senior officer interest contributed to the lost opportunity of a proposed strategic-level Powerful Owl operation. By January 1944, the commander of the 1943 Singapore raid, British officer Lieutenant Colonel Ivan Lyon, had developed a larger-scale plan called Operation Hornbill, a series of SRD operations in the South China Sea area targeting the ports of Singapore, Saigon, Hong Kong and inserting groups of agents to gather intelligence, enable resistance movements and conduct 'small, annoying pin-prick raids' against airfields and railway infrastructure. Churchill would have been proud.

Operation Hornbill depended on the construction of 'Country Craft', a fleet of specially constructed vessels and modified trawlers called Snake Boats, identical to junks operating at that time in the South China Sea. Lyon's vision was for a fleet of Snake Boats, each carrying Motorised Submersible Canoes (MSC), called Sleeping Beauties, that carried limpet mines able to sink two 10, 000-ton ships if correctly placed, to deploy off harbours before launching MSC operated by specially trained commandos to sink ships.

The Snake Boats would also deploy small teams to attack aircraft at airfields and disembark special agents to mobilise local groups to sabotage Japanese military infrastructure. Had there been earlier support for special operations, Operation Hornbill could have included newly developed Welman 'one-man submarines' with more explosives and range and 'undetectable' submersible surface craft, called Welfreighters, for inserting two agents at a time and over a tonne of stores.

Operation Hornbill was mayhem in the making. If Churchill had inspired the SOE with the order to 'set Europe ablaze', Hornbill promised disruption and distraction in Southeast Asia. The Japanese would have lost ships, oil refineries, supply depots, airfields and other critical infrastructure. They would have had to tie down forces to protect themselves from hostile locals, spies and raiders. The psychological impact would have been significant.

Trade union stoppages in Australia stalled the construction of sufficient Snake Boats, and lack of high-level support postponed Hornbill until 1945. Seven Snake Boats were built in 1945, and some were employed in a limited way before the war's end in September. All were decommissioned by the end of 1945. In Hornbill's place in 1944 was an improvised, under-resourced Operation Rimau that ended in disaster at Singapore Harbour a year after the successful first raid on Japanese shipping there in 1943 and cost Ivan Lyon, the Hornbill innovator, his life.

Despite a lack of high-level support from General MacArthur and the American and Australian armed services, independent commando companies and SRD 'M' and 'Z' special units conducted over 100 missions behind Japanese lines in Southeast Asia. Most operations came too late to influence the outcome of the war. The mindset for SRD employment was a British-inspired concept of operations: small teams deploying by sea, land, and air covertly gathering intelligence and conducting small-scale raids. Between late 1942 and August 1945, Z Special Units undertook a total of 84 operations across a wide area of Southeast Asia, including New Guinea and New Britain, Timor, Borneo, the Celebes and Moluccas, Singapore and even at the end of the war, Vietnam, then part of what was known as French Indochina.

The formation of Australia's independent commando companies after Mission 104 left Australia in 1941 was influenced by the development of the British Special Air Service Regiment (SASR). The founder, Lieutenant Colonel David Stirling, wrote that SAS was 'based on the principle of the fullest exploitation of surprise and of making the minimum demands on manpower and equipment'. These were attractive principles for a middle power like Australia, which had well-educated people and limited defence expenditure. Hold that thought.

The prototype UK SASR were highly trained to use any means of insertion [sea, land and air] and, according to David Horner, 'continued to expand and operate with outstanding success in North Africa, Italy, Greece and north-west Europe. ... The Australian Independent companies initially conducted SAS-type operations, but their role changed as the war progressed.' They were brigaded into '[three] Cavalry (Commando) Regiments ... in the last year of the war, the eleven Commando Squadrons fought in Borneo, New Guinea and Bougainville. ... They were tough, highly trained and could operate for long periods in isolated areas.' The brigading of special forces units trained for special operations into regiments subordinate to the operational plans of conventional warfare commanders demonstrated the Australian Army's conventional thinking and preferences at the time. Hold that thought.

Despite conventional thinking and lack of higher-level support, Australia's intelligence-led special operations by independent commando companies early in the war and SRD 'M' and 'Z' special units in the later years were innovative and impressive, but certainly not war winners. The SRD evolved into a fourth fighting force, a formidable armed intelligence service with aircraft, small boats, submersibles, and sporadic access to allied submarines and service aircraft. It had 205 officers and 996 other ranks supported by Flight 200, comprised of eight modified Liberator aircraft for parachute insertion with nine crews and 450 ground staff. Special operations capabilities included sabotage and subversive activities, the training of local guerrilla groups, surveillance and the reporting of Japanese shipping and aircraft movements, the coordination of airdrops and the distribution of arms and supplies.

The United States and Australia failed to develop a special operations strategy during the Pacific War. The Japanese were operating in unfamiliar territory among hostile populations in Australia's regional neighbourhood. Australian politicians and their military advisers did not know how to turn this Japanese disadvantage into Australia's strategic advantage through special operations and enabling populations in Southeast Asia to rise against cruel Japanese occupiers.

The Australian Army did not understand or value special operations. There was no significant increase in the tempo of strikes on Japanese lines of communication, such as railways, harbours and airfields, to distract and tie down Japanese forces. Impeded by American control of resources and involvement in approving operations, the objectives of most SRD operations were limited to intelligence gathering. Senior Australian military officers did not contemplate strikes on Japanese headquarters to paralyse command and control capacity and kill or capture senior Japanese officers. The killing of Admiral Isoroku Yamamoto, commander of the Combined Fleet of the Imperial Japanese Navy in 1943 and architect of the Pearl Harbour attack in 1941, had a significant impact on the morale and strategic planning capacity of the Japanese Navy. This one success illustrated the possibilities of intelligence-led strike operations against senior Japanese officers. This strike against a senior Japanese leader was an innovative American operation enabled by signals intelligence. There was no equivalent level of innovation or ingenuity among Australian military planners, commanders or Defence officials.

SRD was a Powerful Owl prototype Phase 3 de-escalation Response Force. An American estimate states that 264 missions were conducted, at the cost of 417 operatives killed or missing [probably inclusive of locals paid to support SRD operations] against 7203 Japanese dead and 1054 allied personnel [pilots and air crew] and endangered expatriates [British, American and Australia citizens living overseas] rescued from enemy-held territory. Australian estimates vary in the number of operations and losses but align with some American statistics. For a loss of 70 SRD operatives [probably only Australians], the evolved 1000-strong SRD killed just over 7,000 Japanese service personnel, took 141 prisoners, and rescued just over 1,000 allied service personnel and civilians from danger, emulating the Powerful Owl's ability to kill prey ten times its body weight.

The SRD did not survive the general rush towards demobilisation at the war's end and was disbanded entirely, closing Australia's 'eyes and ears' in its neighbourhood and ending special operations in Southeast Asia and the Pacific Islands. Post-war fatigue contributed to the SRD's end. Still, it did not have an ongoing job. The SRD was a hammer for specific nails. Those nails were gone. The Cold War was a few years away and would manifest as Chinese Communist-sponsored violent insurgencies in Asia. No Asian power was troubling Australia militarily or unconventionally in what would be later be called the grey zone in the 21st Century.

After participating in the Korean War, the Australian Army adapted to counter-insurgency operations in Malaya and Vietnam – the correct hammer for those nails. No one expected or planned for countering a grey zone campaign in Australia during the first Cold War. Still, the Russians were displaying early grey zone promise. Their agents infiltrated the Australian Communist Party and conducted influence operations in the Australian trade union movement. Consequently, the Australian Labor Party split in the 1950s and 60s, and conservative coalition governments remained in power until 1972, an example of the Law of Unintended Consequences.

Australian Cold War intelligence and special operations

The 'rush to demobilisation' also impacted Australia's intelligence services. After the government disbanded the CSS, there was a massive loss of expertise and continuity. The Commonwealth Investigation Branch (CIB) was reduced and reformed as the Commonwealth Investigation Service (CIS) with an additional handful of staff focussed on monitoring 'security matters in the interest of the country'. The Defence Department, alert to the Soviet Cold War threat, appointed a Controller of Joint Intelligence and established a Joint Intelligence Centre, a Joint Intelligence Bureau and a Signals Intelligence Centre (SIS) that was disguised under the name Defence Signals Bureau (DSB) – Australia still listened in on others. By 1947, these military organisations had connected to their British and American counterparts for a coordinated effort to thwart Russian espionage, subversion, and sabotage. They were unified in their suspicions about CIS competence. The civilian and military intelligence divide, and dislike continued.

Until the end of the Cold War in 1989, Australia's intelligence services dealt with Russian espionage and attempts to subvert through trade unions, the Australian Communist Party, the Australian Labor Party and created 'front' organisations. Terrorism overtook the Cold War contest as the focus of intelligence services in the 1980s, and this focus continues. By the mid-2010s, ASIO and presumably ASIS began to detect the CCP grey zone campaign.

The rise of special forces

Unlike Australia, Britain and the United States did not discard their special operations capabilities at the end of the Second World War. They covertly raised and employed armed intelligence services and special forces during the first Cold War. The United States divided its intelligence services and special operations capabilities into civilian and military organisations. The CIA was formed in 1947 to develop foreign intelligence and special operations capabilities. The FBI adapted to counter Soviet espionage and subversion in the homeland. Both the CIA and FBI were armed intelligence agencies capable of special operations. After disbanding its special forces in 1945, the US Army raised the 10th Special Forces Group in 1952. The British 22 SAS Regiment consolidated in 1952 for counterinsurgency operations in Malaya.

There was little interest in Australia for enhancing intelligence services and special operations capabilities, though there was British encouragement for the latter. The formation of ASIO had its roots in recognising that Australia needed 'its own counter-espionage organisation. According to David Horner, it also needed to cooperate with and retain the confidence of similar agencies in allied countries. The other motive was the perceived ineffectiveness and incompetence of the CIS. 'It was this inadequacy that led to the formation of ASIO' in 1949 after allies revealed in 1948 undiscovered Russian spies operating in Australia and Australian citizens working for the Australian Government spying for the Soviet Union. Britain and the United States cut Australia off from allied signals intelligence. The Australian government's naivety and complacency were embarrassing, but things were about to change.

In 1952, prompted by the growing Russian threat, Prime Minister

Robert Menzies founded ASIS by charter secretly to collect intelligence overseas and collaborate with allied counterpart organisations. While referring only to ASIO directly because ASIS was still secret, Menzies 'described the Organisation as the 'fourth arm' of Australian defence, after the Navy, Army and Air Force'. Arguably, the nation's special operations capability should have been the other component for the fourth arm of Australia's defence during the Cold War. Still, Menzies knew about Russian political mischief but did not anticipate escalation to violence. Cyber-attacks and terrorist incidents were decades away. Understandably, Defence and the Army had little interest in reestablishing special forces for the Cold War.

It took until 1955 for the Army to raise two Reserve commando units rather than SAS-type units for special operations. Before then, the three services were not interested in re-establishing special forces. After a review in 1957, the Army Minister announced that 'an SAS Company, a form of Commando group' would be raised as part of a new Regular Army brigade group.

It would take until 1999–2000 and the aftermath of 9/11 in 2001 for the SAS Regiment to become a 'force of choice in handling difficult and delicate situations' outside the capabilities of the services. Australian special forces were about to join with allied counterparts in Iraq and Afghanistan in the grey zone when 'war is not war'. For campaigns in the Middle East, the Army enhanced special operations capabilities with commandos. By this time, the ADF had established Special Operations Command, a component of Headquarters Joint Operations Command, commanded by a Major General with appropriately secret links to the CDF, PM&C and the NSC to facilitate limited covert and clandestine action.

After 9/11 and until the overthrow of the Taliban in Afghanistan in 2002, special forces, enabled by strategic intelligence collection agencies and specialised technology, were Australia's Powerful Owl force fighting in the grey zone successfully. After their return to Afghanistan in 2003, Special Operations Task Groups (SOTG), comprised of SAS squadrons and commando companies who did not get along with each other, adapted to counter-insurgency operations. It is debatable how successful they were over subsequent years in this dangerous environment far from home. In 2014–15 in Syria and Iraq against ISIS (Islamic State of Iraq and Syria), new SOTG continued Australia's contribution to the War on Terror.

In the 2020s, the US-led special forces coalition against ISIS, called Operation Inherent Resolve, demonstrates that new global coalitions prefer to employ special forces in an enabling role with local partners rather than bear the brunt of combat operations against sub-national groups like ISIS with their armies.

Lessons for Australia's 21st Century Fourth Armed Service

Looking back to the 1930s and then the conduct of the Pacific War 1941–45, Australia has a history that should not be repeated. There was sufficient warning of the coming conflict with Japan, but Australia depended on others to think strategically. Australia did not develop a complementary self-reliant strategy despite the possibility of a Japanese southern thrust. There was no sovereign hedging strategy against the dominant British Singapore Strategy. Australia had no deterrent strategy or military force in the 1930s to cause Japan, a trading partner, to have second thoughts about attacking Australian sovereign territory after capturing Singapore.

Fixated on allied promises, Australia did not make sovereign preparations until too late. There was no strategy to analyse and deny Japan's stepping stones for invasion – the forward bases Japanese armed forces would need in Southeast Asia. Finding and interfering with invasion forces in their home or forward bases was unthinkable and, if considered, was most likely deemed 'missions impossible'. Australia's human capital was invested in North Africa, Europe and the Middle East, not its regional neighbourhood. Britain recruited and trained Australian pilots and aircrew and benefited from the fighting spirit of Australia's infantry divisions. The Australian government lacked the strategic imagination and initiative to marshal Australia's human capital and use it to defend Australia and its national interests in its regional neighbourhood.

During the Pacific War, Australia ignored the possibility of special operations against Japan's extended supply lines and forces based in Southeast Asia, located far from their homeland. The three Australian service chiefs and American senior officers resented supporting special operations, preferring conventional warfare against Japan's Navy, Army and Air Force, echoing the Napoleonic idea of meeting on a battlefield or at sea at the expected time after discovery.

A combination of Julius Caesar and Aldous Huxley's adages that 'Experience is the teacher of all things, but only the teachable learn from experience' summarises the lost special operations opportunities of the Pacific War. Australian politicians and senior military leaders were not teachable or imaginative. American imagination, US military will, and weight defeated Japan. Australia followed American orders.

Australia started developing a fourth fighting force to hurt the Japanese war effort too late. SOA, formed in 1942, grew and operated as the SRD until the war's end in 1945, despite the service chiefs' hesitation to redirect resources and reluctance to release ships, submarines, aircraft, or personnel for special operations. In 1947, the new Australian Regular Army did not include special forces until a SAS company was formed in 1953/54. This preference suggests that the service chiefs should not sponsor the development of a 21st-century Response Force in the grey zone. Their persistent preference for their services, conventional warfare and protecting and increasing Defence budget allocations suggests that a fourth armed service will need sponsorship from the Prime Minister and separate funding. The challenge will be to find sufficient funds to persuade the services to develop and operate assigned RF maritime vessels, submarines, vehicles and aircraft, as well as specialist capabilities for Phase 3 de-escalation operations against ports, railways, airfields and infrastructure.

Geography and the needs of an invading Asian power have not changed since the end of the Pacific War, but technology has changed astronomically. Like the Japanese nationalists, the CCP must contemplate deploying the Chinese navy, army and air force across the sea and via bases and supply lines along island chains and the homelands of North and Southeast Asian neighbours. The build-up to the First Island chain in the South China Sea is noticeable and well-reported. The CCP can project military force further but will need ports, bases, airfields, and longer supply lines. The CCP grey zone campaign is already 'capturing' forward bases in Australia's near region through negotiation, leasing, and offering lucrative partnerships in the Belt and Road Initiative – no need for a military contest.

Excusing the mixed metaphor, Australia should know its backyard like the back of its hand. A fully developed SOA/SRD was too little and too late to impact the war's outcome with Japan. Could an equivalent technologically enhanced SOA/SRD deter an escalating grey zone campaign? And if deterrence and de-escalation operations fail, could the new SOA/SRD – a fourth armed service called Response Force – become a sovereign Australian game-changer to counter hybrid warfare and, if the CCP rolls the Japanese dice, an invasion through the island chains to the north and northwest?

A Let's Trade, not Argue strategy means the CCP would not detect Australia's RF operators, who will stare and rehearse paced and well-calibrated responses to the CCP's grey zone escalation and watch for hybrid warfare and invasion preparations. Would the CCP have second thoughts about escalating in the grey zone if they knew that every port, land base and airfield was unsafe and supply lines from the homeland were vulnerable? Would the appetite for hybrid warfare diminish knowing North and Southeast Asian governments had a strategy and the capabilities to pounce on grey zone combatants? Would the appetite for bullying Australia diminish if the CCP knew that the price of escalation would be uncomfortable 'stings' and 'jabs' in ports, railways, airfields and infrastructure that illustrated Australia's preparedness and willingness to hurt China's economy as it has hurt Australia's? Would the CCP listen to the messages, 'Please respect Australia's sovereignty. Let's trade, not argue'.

Australia must deter the CCP from escalating by developing 'anywhere and anytime' capabilities to hurt, alert and hurt again. If hurting and alerting are insufficient to prevent escalation, Australia must demonstrate that it can stare at and pace its responses to create more deterrence. If 'going toe to toe' in the grey zone is insufficient, then CCP planning and preparations for hybrid warfare will incur more hurtful consequences.

Creating a fourth armed service takes years, not months, and should not be investigated and acted on only after strategic surprise.

Specialist technology and training bases with optimum curricula, instructors, resources and training infrastructure for RF operations take time to develop. It took 18 months to establish a range of SOE/SRD training centres, technical support schools, and other infrastructure in Australia. The postponement of Operation Hornbill demonstrated both a lost opportunity and the fact that specialist resources take time to develop. 'Country Crafts' or Snake Boats were not built in time to support Hornbill. It took two years for British inventors to develop motor submersible canoes to transport Z Force operators into harbours and ports covertly.

SOE/SRD's difficulties securing resources for operations suggest that a 21st-century equivalent RF must have means for rapid, covert movement and resupply. SOE/SRD had to rely on US submarines and service aircraft. Several special operations were cancelled because submarines or aircraft were unavailable. The Australian Air Force finally established a secret and separate unit, No. 200 Flight, to support special operations in June 1944. Still, it was not fully operational until March 1945, six months before the war's end. The Australian Navy reluctantly enhanced its small boats unit to train SOE/SRD personnel in 1944. Once again, it was too little and too late.

Conclusion

Delaying the development of an RF, a 21st-century SOE/SRD, until after strategic surprise is a risk Australia cannot afford to take. The country urgently needs to align its intelligence services to detect and monitor emerging CCP grey zone threats at home and abroad. Once identified, the Australian Government should deter further escalation by making it clear to the CCP that an RF is in place to discourage any return to bullying tactics.

Strategies for preventing war are not acts of war themselves. The 'Let's Trade, not Argue' approach has significant merit and can be a game changer. Diplomacy backed by a robust RF demonstrates a level of resolve that goes beyond complaint, and a sovereignty that is not solely reliant on powerful allies. For a middle power like Australia, the creation of a fourth armed service designed to detect, monitor, warn, deter, and de-escalate in the grey zone is both affordable and viable.

The development of Australia's intelligence services and special operations organisations during the Pacific War serves as a testament to the country's smart and innovative human capital. Australians were adept codebreakers and skilled mayhem makers. The combination of human intelligence [our spies and those we pay to spy] and signals intelligence [information from other people's telecommunications] yielded impressive results. The volunteer coast watchers around the Australian coast and in the region's islands near the north of the continent were one of the most effective and reliable human intelligence systems that operated during the Pacific War. This history of success demonstrates that Australia has the capability to replicate such achievements in the present day.

The SOE/SRD conducted extraordinary operations that metaphorically killed prey more than ten times their body weight. Investments in Z and M forces achieved amazing returns. Though brigaded by conventionally minded Army commanders for the final operations of the Pacific War in 1944–45, Australian commando companies were comparable to any special forces in the world, making the minimum demands on human resources and equipment but achieving 'the fullest exploitation of surprise'. Australia did it then and can do it now.

CHAPTER 6

Mobilising Human Capital for the Grey Zone

Introduction

The emerging 21st Century Cold War obligates a middle power like Australia to optimise the talents of its people – human capital -, technology and other resources for national security, especially in response to coercive grey zone campaigns that bypass navies, armies and air forces. There is no point in increasing Australia's Navy, Army and Air Force to counter the CCP grey zone campaign described in earlier chapters. They are the wrong hammers for grey zone nails. The ADF does not train for counter-grey zone operations. It exists for conventional warfare and should have the time and resources to rehearse conventional warfighting alongside allies.

Australia's human capital for grey zone defence is not in the ADF, except for special forces and intelligence operatives who have proved adept at counterterrorism and counterinsurgency operations in the Middle East. The bulk of human capital for special operations, such as counter-cyber, information, intelligence, surveillance, drone, strike, and space, is in Australia's corporate and public sectors, not military uniform.

The challenge is to mobilise more Australian talent from its population for the grey zone because this is the battlespace the CCP has chosen and where they are needed. The Department of Defence and the three services will not redirect human capital to this unfamiliar, elusive, complex contest. Furthermore, the ADF struggles to achieve recruitment targets for conventional operations. Efforts to relax recruitment requirements to recruit Pacific Islanders and foreign nationals illustrate this point.

Australians will not volunteer for part-time military service in the 21st Century without a new vision. They will not rally to support their full-time ADF compatriots now and again or be satisfied with a standby mobilisation role to defend against invasion. The 2024 Strategic Reserves Review does not offer a new vision.

In this chapter, the book makes a case for a new vision for Australian part-time military service that marshals exceptional Australians to counter the CCP grey zone campaign, especially in anticipation of escalation to 'dark grey' disruption of ICT systems, supply chains and essential services that will displace and distress Australian communities nationwide. We argue for part-time national service that puts Australians on the grey zone frontline now rather than preparing, figuratively, to dig in along Bondi Beach at some time in the future to face an invasion. Specialised human capital investment in the new Response Force Reserves (RFR) will yield deterrence, agility and proficiency in the grey zone. If that deterrence fails, the high-tech, sophisticated full-time and part-time RF components described in the following chapters will give Australia an additional national security asset to prevent war and enhance national resilience and potency in wartime.

The Reserves back story

Let's review where Australian part-time military service has been to inform where it should go. Australian armed forces have always had full-time and part-time components, which we'll call Regulars and Reserves. Before and since the Federation of the Australian Colonies in 1901, citizens served part-time in the Reserves under different organisational names. Until the Federal Government established the Australian Regular Army (hereafter the Regulars) in 1947, the part-time component of the Australian Army was figuratively, 'the biggest and the best'. Initially called the Militia and then the Citizen Military Forces (CMF), the Reserves were more numerous and influential than the Regulars.

The Reserves' role was homeland defence. Legislation prohibited sending them to fight overseas in other peoples' wars. The traditional vision for the Reserves originated in insecure British colonial times when other European colonial powers sought opportunities on the Australian continent and in Australia's neighbourhood. British regiments served in Australia, and the British Navy was formidable.

The territorial defence of Australia against predatory colonial powers who were Britain's enemies or rivals in Europe was vested in colonists defending their land, economy and way of life. Australian militia units in the 19th and early 20th centuries dressed like and behaved like British Army units but developed some unique Australian 'bush' characteristics, especially during and after Australia's participation in the Boer War 1899–1902.

While the bedrock of part-time military service was homeland defence, Reservists and their governments supported the British Empire militarily in the first half of the 20th Century. Reservists were Australian patriots, but their identity and cultural allegiances were British. Thousands of Reservists volunteered for the First and Second Australian Imperial Forces (AIF) to serve overseas in the First and Second World Wars. The British high command trained and employed Australian expeditionary forces for the First World War and for Australia's participation in campaigns in the Middle East, North Africa and Europe in the Second World War. After the debacle of British command over Australians before the surrender in Singapore, the Americans commanded the returned Second AIF and mobilised the Australian militia for the Pacific War. Australian officers commanded their compatriots at the operational and tactical levels.

As explained in the previous chapter, Australia's sovereign intelligence and special operations capabilities grew during the Pacific War. Talented Australians, some with no military training, broke codes, gathered intelligence and applied technology, and others who were physically fit and adventurous, trained under pressure for special operations. Australia showed it had the human capital to excel in modern conventional warfare and innovate for unconventional warfare. After the Pacific War, the government discarded Australia's unique and practical intelligence and special operations capabilities, possibly expecting that an Asian trading partner would not 'go rogue' again and seek domination of Australia and its neighbourhood.

From the end of the Second World War until 1999, the Militia renamed CMF and then the Army Reserve in the 1970s, remained on standby for homeland defence against invasion as they had done since colonial times. The enduring vision was for them to defend the Australian homeland alongside the Regulars, but Reserve numbers declined without the foreseeable threat of invasion. During the post-Second World War period, the numbers of Regulars increased, and they served in the Korean War, supplemented by a volunteer recruitment scheme, and in Southeast Asia. Australia's participation in the Vietnam War left the Reserves at home again. The Menzies Government decided on a National Service scheme to raise and 'round out' Regular Army units for tours of duty in South Vietnam. Reserve numbers declined, headquarters and units were disbanded, and the remaining units amalgamated.

Australia's participation in the Vietnam War 1962–75 ended the Reserves' supremacy in the Army. Though Australia's first strategic priority was, and still is, defending its homeland, the Regulars' overseas experience and expertise derived from full-time employment with modern weapons overshadowed the Reserves for national security. Without a threat to the homeland, the Reserves were on standby without meaningful, exciting things to do except train like the Regulars, always falling short of the Regulars' proficiency. The Reserves were mocked as 'weekend warriors' and given other names such as koalas, meaning slow, cuddly marsupials unable to be exported, i.e. sent overseas to fight, and 'chocos', meaning fragile chocolate soldiers that would melt under heat, a metaphor for lacking a fighting character and grit. There were questions about the usefulness of the Reserves, a historical legacy, not a national asset for Australia's contemporary defence.

After the end of Australia's participation in the Vietnam War in 1972, the Regulars received priority for resources in tight Defence budgets. Reserve morale hit rock bottom over the next 25 years until the unexpected Australian-led multi-national intervention into East Timor in September 1999, followed by further interventions into Solomon Islands and deployments to the Middle East in the first two decades of the 21st Century. The Regulars could not meet these commitments without reinforcement. A Vietnam War-era conscription scheme was not seriously contemplated. The government and the ADF turned to the Reserves to ease pressure.

Despite their reduced numbers and low morale, Reservists rallied and volunteered to serve with the Regulars on overseas operations. The Reserves began training and deploying garrison troops to East Timor [Timor Leste after 2002] and Solomon Islands. Reservists grew to become 18 percent of Australian military personnel serving overseas. Numbers declined again after 2006 when the operational tempo eased, and Reservists were no longer needed. Despite prominent and successful contemporary employment at home and abroad, Reserve numbers declined to an all-time low in 2016 and remained low.

The Reserves rallied again to support the nation's responses to the COVID-19 pandemic and catastrophic bushfires and floods in 2019–2021. In 2020, the Morrison government called out thousands of Reservists to support emergency services, community organisations, and police responding to the Black Summer Bushfires and significant flooding events. Reserve depots nationwide became hubs for emergency service responses and shelter for displaced and distressed people. Soon after the bushfires and floods, Reservists deployed nationwide to command and conduct domestic operations in support of State and Federal authorities during the COVID pandemic in 2020–2022. Significantly, Reserve headquarters rallied to command and employ task forces comprised of Regulars and Reserves for these national efforts. The Reserves became a national asset to back up Australia's State and Territory first responders, with a new, unexpected role as the ADF's force of choice for domestic emergency response operations. These operations were practical examples of the Reserves employed as a response force for national resilience. The die was cast out of necessity and convenience for the Regulars' who prefer not to become involved in natural disaster responses. Hold that thought.

Once again, when the operational tempo and meaningful employment declined, Reserve numbers plummeted. The traditional vision of part-time military service was the cause of the low numbers and poor morale. Defence White papers from the 1970s and numerous institutional reviews into the Reserves specified that they were part-trained Regulars standing by for mobilisation to defend the homeland from invasion or expedient call out to back up the Regulars if the operational tempo put them under pressure. They would receive further training, modern weapons and equipment only if an enemy invasion fleet threatened. By 2024, with COVID pandemics, bushfires, and flooding disasters in 'the rearview mirror', the Reserves returned to being 'weekend warriors', playing at being Regulars with minimal resources and constrained paid training time.

The Reserves wither when they have no meaningful role in the nation's defence. The 2024 National Defence Strategy (NDS2024) is silent about a role for the Reserves but called for another review. The 2024 Strategic Review of the ADF Reserves fails like every review since 1970s to specify a meaningful role for the Reserves.

Observations

In 2024, Defence and the Army, and to a lesser degree the Navy and Air Force, 'dropped the ball' on marshalling Australia's human capital for part-time military service for national defence. Slogans like 'one Army' and 'Total Force' and calls for a two-division Army – one full-time and one part-time – are hollow and futile when NDS2024 does not give the Army Reserve a real job. Since the end of Australia's participation in the Vietnam War in 1972 and the publication of the first Defence White Paper in 1976, Defence's strategic guidance in white papers and strategic updates, culminating in the NDS2024 silence, leaves the Reserves chewing up an estimated $400million annually with no capability and collective role to play in Australia's defence.

Over the first decades of the 21st Century, fewer Australians have offered their time and talent to voluntary part-time military service. The government and Defence's ambivalent treatment of the Army Reserve, cumbersome centralised recruitment processes, and token roles contributed to this failure to marshal Australia's human capital for its defence. Failed government policy and poor Defence oversight are responsible for the Army Reserve becoming an under-resourced, under-trained and low-morale militia, despite impressive performances in the 21st Century backing up the Regulars overseas and becoming a force of choice for domestic operations, especially in response to natural disasters.

Continuing to align Army Reserve training to traditional

occupations, such as infantry, armour, artillery, engineers, transport, signals, intelligence and ordnance, has been and continues to be token and uninspiring. Part-time training to achieve low-level expertise is sub-optimal. Talented Australians who are needed to defend the nation are being left with the choice to forego civilian employment and join the Regulars or climb into uniform and serve in units that have historical identities and proud traditions but whose last specified role is to guard infrastructure in northern Australia while the Regulars look out over the air-sea gap, hopefully with enough missiles to fire at an invasion fleet.

It is time to reimagine part-time military service in Australia and marshal Australia's human capital for service in the nation's fourth armed service: the Response Force Reserves (RFR).

The RFR and the Second Cold War
Strategic guidance and commentary in 2020–2024 slashed strategic warning time, comfortably set at ten years in the latter part of the 20th Century, to zero, evoking the 1930s before the Second World War. Australia has human capital ranked amongst the highest in the world, which governments should marshal urgently and astutely to enhance national security in these troubled times. It is time to look at the Reserves through the lens of national resilience and countering grey zone campaigns rather than anchoring them in a pre-invasion mobilisation role that withers them on the vine. Australia must recruit more citizens for part-time military service for national resilience and grey zone defence – no longer traditional voluntary military service but modern voluntary national service.

It is time for a human capital rather than a traditional military mindset that 'more troops is better'. It supersedes the idea that Reservists are partially trained Regulars. This new focus prioritises attracting recruits with civilian skills for national resilience first and for traditional military occupations second. The RFR will not emulate Regular occupational groups, functions and organisations. The new vision is for national resilience, so organisational design and resourcing, personnel skill sets, training programs, contingency rehearsals and operations should first focus on mitigating natural and grey zone-inflicted disasters because these calamities are more likely than invasion. The new RFR will be frontline and online specialist operators and extraordinary operations RF warriors, not 'weekend warriors'. No more parade ground drill, occasional shoots at a rifle range, one weekend a month camping or finding something to do at the depot on Tuesday nights and a two-week annual camp.

The 21st-century economy and service industries have spawned valuable corporate workforces for employment in the grey zone. Australia can use the RFR to access these skilled personnel and workforces that the private and public sectors select, train, manage and employ day-to-day. Under new recruitment and part-time service arrangements, the government can counter an escalating grey zone campaign by identifying and rehearsing latent capabilities in the corporate sector. An RFR embedded in the public service and private business sectors and maintaining close community connections optimises Australian human capital for internal security, resilience and homeland defence while giving full-time RF units access to high-quality personnel for sophisticated extraordinary de-escalation operations.

The logical starting point is a new strategic vision for part-time service for national security in the 21st Century. The new vision involves defending Australia in the grey zone while being mindful of escalation to regional conflict. The vision must apply Australia's competitive advantage in human capital for three modes of service. Firstly, high readiness employment of RFR civilian specialists to counter the CCP grey zone campaign anywhere and at any time. Secondly, for national resilience before and after natural and grey zone-inflicted disasters, and thirdly, in readiness for conventional conflict to defend the homeland should Australia's grey zone and conventional deterrence strategies fail and war be imminent.

These three service modes underpin 'end-to-end' strategic messaging that starts with a 'Let's trade, not argue' message, continues after escalation with a 'There are consequences for bad behaviour' message and ends when hybrid warfare looms, with a 'Look what you made us do' final message.

The vision

The vision is for the RFR to marshal Australian human capital for voluntary part-time national service that bolsters national resilience against foreign coercion and natural disaster impacts and defends Australia's territorial sovereignty. The vision does not change the bedrock mission of part-time military service, namely, the defence of the homeland and Australian communities. It reorders four complementary tasks for voluntary part-time service: countering the CCP grey zone campaign, enhancing the Federal Government's disaster response capability, ensuring homeland resilience if the CCP escalates its campaign to violent disruption of the Australian economy and essential services, and bolstering territorial defence during war.

These compelling tasks will motivate Australian citizens to protect their nation part-time against threats to life, property, and community while remaining in civilian employment. These tasks require selection, induction, training and management systems that optimise individual citizens and civilian workforce groups rehearsing contingencies to counter a threatening grey zone campaign and harden Australian resilience in the aftermath of natural disasters and escalating grey zone attacks, especially cyber-attacks that will also displace and distress communities.

Command and control

The Army should not raise, train, sustain, or employ the RFR as a revamped Army Reserve. Defence and the Army have never been and will never be structured or directed to conduct Phase 1 deterrent diplomacy backed by an armed intelligence service, Phase 2 de-escalation operations or Phase 3 extraordinary operations. NDS2024 locks Defence and Australia's three armed services into conventional defence against invasion – full stop – and governments from both sides of Australian politics have and continue to promise extraordinary amounts of money to fulfil this mission. For Defence, the Army, and Australia's growing Defence industry and consultancy services, the grey zone is a Red Herring – a distraction from core business.

The RF's homeland, regional and international missions mentioned in Chapter 4 and to be specified in Part 3 mean PM&C must command the RFR at the strategic level, not Defence. The commander with ultimate responsibility for RFR performance should be the Chief of National Security (CNS), not the Chief of the Defence Force (CDF). Response Force Headquarters (HQRF) in PM&C should be responsible for raising, training and sustaining the RFR, not Defence. The vision and tasks of the RFR align with Home Affairs and the ASIO/AFP/ABF component of the RF, not with Defence. The RFR marshals human capital for the grey zone frontline, not for the DSR2023 Army Reserve's task of protecting infrastructure in northern Australia while the Regulars defend against an invasion. The grey zone frontline has no borders, so RFR personnel will operate within and beyond Australia's borders, primarily online but physically anywhere and anytime when called up to do so.

Why propose such a radical change? Surely, this transfer of budget allocation and assets from Defence to PM&C, DFAT and Home Affairs is 'breaking too many eggs to make the omelette'? Defence will oppose losing $400 million and repurposing the Army Reserve as the RFR. Yes, these changes will break some eggs, but they will be the eggs of self-interest and adherence to tradition. The Army Reserve and the RFR can co-exist because they have different missions, but the Australian taxpayer should not foot the bill for both. The RFR replaces the Army Reserve because it will still be capable of physical homeland defence.

If our prognosis for the Army Reserve, assured by the NDS2024 silence and another efficiency review, is correct, the Army Reserve will wither on the vine anyway. The 2024 Strategic Reserves Review reflects all previous efficiency reviews, hollow units and persistent problems with recruitment and retention. With a nod to the grey zone, it recommends   recruiting cyber and space domain specialists but does not explain how to do so.

It is the government's job to decide whether an annual expenditure of approximately $400 million is value for money for the Army Reserve in its present condition, not this book's authors. The government can remove the Army Reserve from Defence and repurpose, rebadge and refresh units and infrastructure into the RFR. The alternative is to find new money to establish the RFR and allow the Army to continue its stewardship of the $400 million annual outlay for the Army Reserve.

Let Australians decide whether to wear an Army uniform as a 'weekend warrior' or an RFR uniform as a grey zone national resilience warrior. When Australians shift their preferences to the RFR because its purpose is to keep Australia safe now, not defend against invasion in the future, the government will redirect funds from hollow Army Reserve units and neglected depots to new RFR units and refurbished and new depots. Many Reservists will want to transfer their time and talents to the RFR because it protects communities and, for the most talented, is on the frontline of a threat to Australia's well-being and way of life.

Conclusion

Whether ignorant or dismissive of the grey zone threat, military traditionalists will argue against the government expanding part-time military service to natural and grey zone disaster responses. They will fight hard to retain historical unit identities and the status quo of dressing up and training for conventional military activities in good company. They will object to infantry, artillery, armoured, engineer, transport and ordnance units being retitled, rebadged and repurposed for grey zone operations and national resilience. These traditionalists are the equivalent of those who fought hard to retain the Polish cavalry's identity, traditions, horses, and training regimes before the German tanks rolled across the border in 1939.

The Regulars exist for conventional warfare and should have the time and resources to rehearse conventional warfighting alongside allies. The RFR can and will support the Regulars, especially with specialists, but no longer from structures, training regimes and pooled resources that maintain the low readiness levels of a semi-skilled stand-by Army Reserve.

The government must turn to Australia's corporate sector to recruit and foster the RFR. The corporate sector – the engine room of Australia's economy where trade is so important – has so much to lose if the proposed Let's Trade, not Argue strategy is not implemented. According to Prime Minister Albanese, one in four jobs in Australia comes from trade. Australian businesses know how trade embargoes and cyber-attacks hurt. Corporations will rally behind the book's strategy and proposed full-time and part-time Response Force because its emphasis is not on military defence. Its focus is on defending a trading relationship that suffers when the CCP bullies with trade embargos, tariffs and cyber-attacks.

The RFR will not be comprised of semi-skilled, under-trained Regulars; it will be fully trained and capable of tasks that leverage civilian skills and apply to real-world contingencies in the homeland. In these dangerous times, the Reserves must transition from a historical legacy to a valuable RFR national asset – an asset backed by the corporate sector that Australians will be attracted to and proud to serve.

How will a government-corporate partnership in the grey zone work? Large and small businesses – the corporate sector – constantly recruit, train, and employ people. Recruitment and training systems are well-established and adapt continuously to changing market forces. Businesses compete for the best people to contribute to their commercial success. The government must explain the RFR's vital mission and offer incentives to the corporate sector to encourage their best people to join the RFR and for RFR volunteers to have flexible employment arrangements that allow them to train and respond when required. These employees will benefit from RFR remuneration and the personal and professional development the RFR offers. This enhancement of skills, knowledge and attributes adds to these employees' contribution to their employer's commercial success.

Finally, disruption, coercion and hybrid warfare will come onshore well before an invasion fleet, missiles, drones, and bombers trouble the nation. The reimagined and established RFR, a partnership between the government and the corporate sector, is essential to meet urgent and alarming challenges for Australia's national security emerging in the 21st Century. The best of Australia's human capital must volunteer for part-time national security service to defend in the grey zone and respond to natural and grey-zone-made disasters.

Part Three – The Response Force

Where will the Response Force fit in? What will it do?

Chapter 7 describes how the Response Force (RF) will fit into Australia's national security arrangements. The Army Reserve will transition to become the Response Force Reserves (RFR) with a new vision for contributing to grey zone defence and national resilience. The Prime Minister will have overall command through a Chief of National Security in PM&C. Home Affairs and Foreign Affairs will employ assigned RF and RFR teams nationally and internationally.

Chapter 8 describes what the RF and RFR will do nationally, regionally, and internationally. For Phase 1 and 2 grey zone defence, the RF and RFR will conduct mainly informational and counter-cyber operations and natural disaster responses when required. For Phase 3, an armed RF component will conduct extraordinary operations to achieve specific strategic effects.

Chapter 9 describes the steps needed when the Australian people and their government realise that the CCP is escalating in the grey zone and will not stop until they are deterred from doing so.

CHAPTER 7

Where will the Response Force Fit In?

Introduction

Deterrence is impossible without a Response Force (RF) capable of extraordinary operations anywhere and anytime in the Australian homeland, near region and worldwide. Australia is not ready for persuasive Phase 1 activities escalating to coercive Phase 2 and 3 disruptions. Tibet, Hong Kong, the islands and atolls of the South China Sea and assorted ports and airfields are now under CCP control because there was nothing to deter Phase 1 CCP takeovers.

Australia must be ready to de-escalate with an RF during an escalation to Phase 2 that may begin as soon as 2025. Phase 2 may look and feel like 2021, when the CCP was doing billions of dollars of damage to the Australian economy, increasing the intensity and frequency of espionage and cyber-attacks, and leaked its 14 grievances with Australia (Google China's 14 grievances with Australia).

Creating public awareness, understanding and commitment to the RF and its mission is crucial before the government acts. Defence, Foreign Affairs and Home Affairs know about the grey zone threat, but there is no political incentive, authority or responsibility to act. The CCP courted and did not coerce, the Albanese government in 2024. This book predicts a return to bullying or worse in 2025/26 (see Chapter 2). It can and should inform government policy when 'reality bites' but will never drive government policy and resource reallocation until the Australian government and people have moved from being alert to the danger to becoming alarmed when they experience coercion and disruption.

Public opinion matters. The government took legislative action by 2018 when public opinion peaked at 80 per cent of respondents to the annual Lowy Institute Poll, assessing China to be more of a security threat to Australia than an economic partner. In 2023, over

70 per cent of respondents still assessed China as a possible military threat to Australia. By 2024, just over 50 per cent of respondents, possibly lulled by the CCP courtship of the Albanese government or just in hope, see China as more of an economic partner than a security threat. The 'time is ripe' for convincing the Australian people that a Let's Trade, not Argue' message to China backed by an Australian RF should complement a 'submarine and missile' message supported by the United States.

Let's begin Part 3 by discussing the implementation challenges and where the RF will fit into Australia's national security arrangements. Then, in Chapter 8, we will discuss what the RF will do and what it will need to fulfil its mission. Chapter 9 will chart the way ahead – what are the next steps?

The implementation challenges

In addition to capitalising on the teachable time when Australian attitudes move from being alert to alarmed, the Let's Trade, not Argue strategy will face several significant implementation challenges. It is politically, ethically, and morally uncomfortable for a liberal democracy like Australia to 'harden up' for the grey zone. Sections of the population, political parties and civil society are wary of secret organisations. Some will characterise the proposed armed component of Response Force as an Australian CIA, Mossad or MI6 despite its origins in Australia's military history, not any other nation's history, precedence or example.

It is even more uncomfortable to plan for and maintain a presence in someone else's homeland when an adversary, like the CCP, does not respect the sovereignty of other nations and has grey zone operatives located in Australia and throughout its near region. It would be easier if a state of war permitted wartime legislation and the employment of conventional military forces. But that has not and will not happen in the 21st Century if the first 20 years are instructive.

In 2025, no government department or agency wanted the RF or to be forced to set up, host and employ it. An RF does not fit neatly into any government portfolio's core business or current priorities, and there is no money in the budget forecasts for an additional national security strategy. Defence, Home Affairs, Foreign Affairs, Attorney General's and the NIC will resist the RF and RFR because a de-escalation strategy will re-direct some of their funds and organisational efforts. Still, Foreign Affairs and ASIS will appreciate deterrent diplomacy backed by an RF to gain the CCP's respectful, albeit resentful, attention and acceptance of Australia's political autonomy and alliance preferences.

Consensus for an RF must come from the top. The Prime Minister and his department must take the lead to develop a shared understanding of the grey zone threat and consensus among departments and agencies at Federal, State and local levels, and the corporate sector that a de-escalation strategy backed by an RF can meet it and restore a trade-based China-Australia relationship.

The last time an Australian government received and adopted an externally developed de-escalation strategy for protecting national interests in its regional neighbourhood was in 2003. This 2003 precedent proves that a Prime Minister can turn around consolidated departmental opposition without needing a favourable opinion poll. When the big departments obstruct against the book's proposed strategy, will Anthony Albanese or his successor find the courage John Howard displayed in 2003 to conduct an extraordinary whole-of-government and whole-of-region de-escalation intervention?

What happened in 2003?
Dr Elsina Wainwright, then Program Director at the Australian Strategic Policy Institute (ASPI), wrote *Our Failing Neighbour – Australia and the Future of Solomon Islands* in late 2002. She proposed an Australian-led law-and-order intervention in Solomon Islands to de-escalate increasing violence from armed ethnic militia groups who were threatening the government and pushing Solomon Islands to the brink of civil war. She took a draft of her de-escalation strategy and extraordinary intervention to Defence and presumably to Foreign Affairs and possibly Attorney General's in December 2002. The Department of Home Affairs was 14 years away.

Officials in these departments ignored her despite warning about the consequences of Australia not doing enough for Solomon Islands. They turned 'deaf ears' to her arguments that existing policies of advice, development projects and patrol boats were not working. She called for an Australian whole-of-government sponsorship of a Pacific Islands whole-of-region intervention in Solomon Islands that was not neo-colonial but the actions of good neighbours de-escalating armed political violence.

Though not naming it in 2003, Wainwright proposed an extraordinary deterrent operation, not a conventional armed law enforcement or ADF intervention. No one in the departments responsible for protecting Australia's national interests agreed with her. In January 2003, Australia's Foreign Minister, Alexander Downer, was concerned about the neo-colonial look of Australia intervening in Solomon Islands. Why did her recommended intervention occur six months later, in July 2003?

For two months, nothing significant happened. However, in April, Wainwright's paper found support in the Solomon Islands Government. Prompted by a draft sent to the consulate in Canberra for comment, Prime Minister Alan Kemakeza wrote to Prime Minister John Howard on 22 April 2003 seeking Australian intervention. Defence, Foreign Affairs and PM&C advised John Howard to stay out of Solomon Islands in a rare consolidation of departmental opinion. For their part, Defence officials by-passed advice from the Chief of the Defence Force, General Peter Cosgrove, who supported a 'circuit breaking' intervention – a de-escalation mission. Defence officials persuaded Defence Minister Senator Robert Hill to suggest a 'package of targeted measures.'

In the end, John Howard brought Alexander Downer and Robert Hill together. He directed them to have their departments prepare intervention options for the National Security Committee of Cabinet's (NSC) consideration. He had decided it was Australia's responsibility to step up its statecraft and act decisively to de-escalate the Solomon Islands crisis. It was to become an extraordinary deterrence operation that would provide security for a regional police operation to disarm militia groups threatening civil war.

Pre-publication copies of Wainwright's paper were available for policymakers and planners. It provided a rationale, several prescriptions and a plan. Arguably, no one in Defence, Foreign Affairs or Attorney General could have or would have designed a whole-of-government and whole-of-region intervention. Without Wainwright's insights, the intervention should have had General Cosgrove in overall command and the ADF 'upfront' with an AFP contingent and some diplomats in support. These ADF-led command and control arrangements had applied to all but one of the other Keating and Howard government interventions in Australia's near region in the 1990s and early 2000s.

The Regional Assistance Mission to Solomon Islands (RAMSI) turned out to be totally 'arse about' – a diplomat in charge and police up front with soldiers, ships and helicopters behind in support. An Inter-Departmental Task Force [IDTF] succeeded in persuading the NSC to appoint Nick Warner, a senior Australian diplomat in charge of an AFP commissioner, Ben McDevitt, who was in command of assigned AFP, ADF and regional police and military contingents. While a senior ADF officer, Colonel Paul Symon, advised Nick Warner, command of 1,200 ADF personnel serving on the ground and operating ships and helicopters was given to Lieutenant Colonel John Frewen, a battalion commander, to ensure that the ADF was 'outranked' and could not assert undue military influence on how operations were conducted. The IDTF also designed the intervention to be 'whole-of-region' by including contingents from police and military units from New Zealand and other Pacific Islands neighbours, thus *enabling* [authors' emphasis to reference extraordinary operations Chapter 8] the region to de-escalate a regional problem.

There are still debates about RAMSI's duration and expense after it disarmed the militias and restored law and order by 2005. Still, no one doubts its success as a circuit-breaking de-escalation of lawlessness in 2003–04 that has endured, with a brief regression in 2006, for 20 years. The intervention was Prime Minister John Howard's de-escalation masterstroke, albeit a larger-scale extraordinary operation than the calibrated ones envisioned for the RF in Chapter 8.

So what? That was then, and this is now. De-escalating the CCP grey zone campaign differs significantly from de-escalation in Melanesia. The Pacific War occurred more than 80 years ago. Let's 'fast forward' to 2025–26, when the CCP will escalate again. What can be learned about implementation challenges from the innovative RAMSI intervention in

2003 and the Special Operations Australia/Services Reconnaissance Department (SOA/SRD) experience in 1942–45?

Implementation time limits

Australia started to develop the equivalent of Response Power and a Phase 3 RF in 1942, only after a strategic surprise at Pearl Harbour in December 1941 and the fall of Singapore in March 1942. Service chiefs [RAN, Army, RAAF] opposed the diversion of their resources for SOA/ SRD extraordinary operations in 1942 and the following years. By the time SOA/SRD operations and technology were ready to disrupt the Japanese war effort in Southeast Asia, the war was over. It took years – not months – to select, equip, and train the SOA/SRD and to plan, deploy and conduct operations.

It took two years for the Australian armed forces to develop a training and rehearsal system for Z and M Special Forces. The services reluctantly directed resources for SOA/SRD training. After depending on British Mission 104 operatives, Australians proved to be visionary innovators and improvisers. Mission 104 set up a training base and conducted commando training at Wilson's Promontory in southern Victoria in a climate suited for Europe. After they were sent home, Australian innovators set up a base and conducted training at Fraser Island in Queensland in an environment suited to Southeast Asia. Repeating this haphazard shoestring process of the 1940s for the 21st century RF or copying how allies do business ignores history and fails the commonsense Pub Test. Australia's hi-tech companies and the corporate and public sectors have sophisticated training and rehearsal systems, so why not the RF?

Australia cannot achieve deterrence by slowly building the RF in response to CCP escalations. The SOA/SRD development was too slow and came online too late to have an impact during the Pacific War. The RF must be real and ready, even for Phase 1 deterrence. Removing the temptation to escalate to Phase 2 in Phase 1 saves China and Australia a lot of trouble. From the beginning, the RF must train and be equipped for strike operations, not just for lower-level non-violent extraordinary operations to send warnings and 'back-off' messages. Metaphorically, it trains for astute hard punching while knowing how to jab and poke to persuade the CCP to call the grey zone campaign off. Metaphorically, the RF is a guard dog with a muzzle and warning bark, but if ignored, the muzzle comes off and will bite.

Defence Department resistance

Defence and service chiefs will oppose Australian Response Power. They focus on warfighting, not war prevention, retaliation, not reconciliation. They will argue against a fourth and armed service that reallocates resources to the RF for extraordinary operations. Raising an RF with high-technology detection, surveillance, de-escalation capabilities, and maritime, land, air, cyber, and space stingers threatens Defence budget dollars.

Defence and the services will manipulate the strategic debate, as they did in the 1930s before the Pacific War. NDS2024 promises hundreds of billions of dollars for military hard power [submarines, ships and missiles]. It ignores part-time military service. The 2024 Strategic Review of the ADF Reserves fails to give the Reserves a strategic mission. It identifies but has no strategy or plans for countering the CCP grey zone campaign. Defence will support NDS2024 exclusively and argue that a Let's Trade, not Argue strategy wastes time and money and is the responsibility of other departments and its agencies.

The other reason for Defence resistance is that the strategy is about 'special', not conventional operations. Despite the 21st-century employment of ADF special forces in the War on Terror, Defence dislikes the priority special forces receive. Inconveniently, alleged unsavoury special forces' activities in Afghanistan complicate any justification for special forces securing more authority, responsibility, and resources for the grey zone. (Google Australian Special Forces in Afghanistan).

Special Forces and the grey zone?

The ADF special forces will assert themselves as the force of choice for the grey zone, as they have since 9/11. They will highlight their 20 years of overseas experience and guardianship of Australia's counter-terrorism response. They will want to take control of an RF and raise, train, and rehearse it as an additional elite special forces unit, possibly emulating the American CIA, Drug Enforcement

Administration and Delta Force organisations.

The ADF Special Operations Command should not raise, train, and maintain the RF. Current personnel selection practices focus too heavily on martial skills and insufficiently on the skills, knowledge, and attributes necessary for sophisticated grey-zone responses. Their special operations would be too lethal, too early, and too provocative. The ADF special forces do have an undercover strategic reconnaissance role, but 20 years of the War on Terror have diminished it.

Enough smart Australians?

The RF is the hammer for the Phase 1, 2 and 3 grey zone nails. Australia can become the best in the world at extraordinary operations by consummate professionals to achieve specific effects. Still, there will be doubters that Australia has the talent to accomplish this lofty goal – another implementation challenge.

Does Australia have the human capital, ingenuity and organisational imagination to implement an RF-led de-escalation strategy? An analogy would be a New Zealander arguing New Zealand lacks the human capital to field a world-class Rugby Union team. Indeed, other countries with larger populations have more resources, imagination, experience, sophistication, technology, and ingenuity to win the Rugby Union World Cup. The performance of New Zealand's national teams, the All Blacks and Black Ferns, defy this pessimistic assessment. Envious Australians argue that New Zealand's advantage is a national focus on Rugby Union. They argue that Australia has competing team sports that deny Australia's national Rugby Union teams, the Wallabies and the Wallaroos, the elite athletes they need to excel internationally.

These assessments and observations are nonsense. New Zealand Rugby Union selects, trains, manages, and rehearses its human capital with organisational machinery, ingenuity, and commitment, which begins at the community level. With the enthusiastic support of the nation, the New Zealand Rugby Union employs specially selected, trained and managed men and women with a strong ethos and skills for collective success against opposing nations with larger populations and more generic resources.

Response Power depends on expertly selected, trained, and

rehearsed intelligent and capable people – consummate professionals. RF task forces will need intelligence, cross-cultural, sociology, psychology, surveillance, cyber, political, and information operations specialists. Foreign Affairs, Defence, and Home Affairs cannot generate the range of intelligent people necessary to meet Response Power requirements with single department efforts.

Australia's corporate sector will be crucial. The prime minister and his department must marshal Australia's best and brightest for the RF in conjunction with the corporate sector, which already has much of the human capital Australia needs to defend and de-escalate in the grey zone. The public sector outsources ICT, information, technical and other specialists from the corporate sector. The corporate sector recruits, trains and employs the best, and the public sector does so for the rest.

The government must educate Australians about the grey zone and convince corporations and communities to rally. This is a call to peace-making de-escalation, not a traditional call to arms.

RFR implementation challenges

Two implementation challenges face the Response Force Reserves (RFR) – inexperience and unwillingness. Below, we argue for PM&C, a Chief of National Security (CNS) and an HQRF to incorporate the Army Reserve and its infrastructure and establish the RFR. This proposal involves massive organisational change. Currently, the PM and his department coordinate rather than command and facilitate solutions with departments and agencies rather than prescribe strategies and their implementation. Home Affairs has no experience employing the RFR for natural and grey zone disaster relief. Foreign Affairs has no experience in employing part-time RF specialists. The implementation challenge is to change the status quo and overcome inexperience to establish the RFR.

Critics will question whether PM&C and the Home, Foreign Affairs and Attorney General's portfolios have the expertise and experience to manage Australia's new voluntary part-time national service workforce for grey zone defence and then to counter hybrid warfare if de-escalation does not work – a fair enough concern but only up to a point. In 2017, Australia transformed homeland defence by consolidating people, organisations and assets into the Home Affairs portfolio. Following that precedent, the nation can formalise the Prime Minister's responsibility to keep Australia safe in the grey zone. He and his department can partner with departments, agencies and the corporate sector for national and international grey zone defence and consolidate the best people, organisations, and assets into a whole-of-government RF corporate sector partnership.

Defence will be unwilling to set up and employ the RFR. The DSR2023's allocation of an infrastructure protection role in northern Australia, and the NDS2024's silence about part-time military service and the under-whelming 2024 Strategic Reserves' Review confirm that Defence does not comprehend or prioritise 21st Century part-time military service. The battle in Defence to stop repurposing, rebadging and transforming the Army Reserve into the RFR will be fierce. There will be Defence officials and Reservists who will argue that the Army Reserve has a history that goes back to colonial times of emulating conventional military forces rather than a special forces variant like the RF. These institutional defenders and critics will develop a narrative that the RF is an Australian CIA with unsavoury and unethical possibilities or an Australian Mossad.

The alignment of attributes and functions is the strongest argument for the Reserves to transfer away from Defence to PM&C and Home Affairs and, to a lesser extent, to Foreign Affairs to employ part-time specialists. The RFR's new vision is about homeland defence (See Chapter 6), which aligns with the day-to-day business of Home Affairs. Defence's strategic guidance to the Reserves to train on a 'shoestring' awaiting mobilisation against invasion and to 'round out' Regular units when the operational tempo puts them under pressure does not align with homeland defence. One of the Reserve's attributes is that members live and work in communities, and operate from community-based depots. This community connectivity aligns strongly with homeland resilience in responding to natural and grey zone disasters and civil emergencies.

Prime Minister Albanese or his successor must get involved, spend political capital and marshal the corporate sector. This leadership may involve 'banging departmental heads together' as John Howard did in 2003. This book's logic and Mark Armstrong's PhD thesis, *Historical Legacy to National Asset: the Australian Army Reserve in the 21st Century*, provide the justification and evidence for change.

Command and control

Assuming the PM has successfully overcome implementation challenges from Cabinet colleagues and their departments, he must specify command and control, where Response Power and the RF fit in and what the RF will do. The following proposals will help. We try to keep it simple. Sorry about the acronyms.

Response Power is the fourth leg of Australia's defence chair. The other legs are Maritime Power (Navy), Land Power (Army) and Air Power (Air Force). The Defence portfolio manages Australia's maritime, land and air power. The CDF commands the ADF on behalf of the Minister for Defence and advises the government in the NSC on employing the ADF to achieve the desired effects. The CDF delegates responsibility for conducting ADF operations that harness maritime, land and air power to defend Australia and its national interests to a Chief of Joint Operations and his staff at Headquarters Joint Operations Command located at Bungendore, near Canberra.

The NSC is the peak group, informed by the Secretaries of each department, that knows about national security threats and decides on maritime, land and air power employment. The NSC is also responsible for strategic targeting and effects. Logically, the NSC should employ Response Power to deal with the grey zone threat because no single department can or is willing to do so. In 2024, the PM chaired the NSC, comprised of the Deputy Prime Minister and Minister for Defence (Deputy Chair), Treasurer and Attorney General, and the Foreign Affairs, Home Affairs, Finance, Climate Change and Energy, Defence Industry and International Development and the Pacific ministers. Accordingly, the PM should facilitate NSC consensus on the severity of the grey zone threat and the employment of Response Power to counter it.

A Chief of National Security (CNS) should command the RF on behalf of the PM within PM&C and advise the NSC on the RF employment nationally, regionally and internationally to achieve desired specific effects. PM&C has had a CNS before but only empowered them with a small staff and some authority, not an RF with operational responsibilities, accountability, capabilities and resources. Put simply, the CDF advises the NSC on Australia's employment of maritime, land and air power, and the CNS advises on response power. Paul Symon, former Director General of ASIS, argues for the reappointment of a CNS. He writes, 'A renewed emphasis needs to be placed on grand strategy and strategic thinking. In essence, Australia needs to concentrate on and articulate the big picture better. Adam MacAllister, a visiting fellow at ASPI, joins Paul Symon, Ben Scott of the Lowy Institute and the late Jim Molan, former senior ADF officer and Senator, in calling for a National Security Strategy to 'address malign Chinese behaviours, and enable a robust national security dialogue among the public, industry and international partners ... [and] prudent cooperation and facilitating an effective deterrence policy.'

Through PM&C and CNS, the NSC would be responsible for strategic level coordination, and the Home Affairs and Foreign Affairs would assume the responsibility of commanding RF units at the operational and tactical levels. The CNS raises, trains and sustains the RF through an HQRF located in PM&C but will assign RF units to work under the operational control of Home Affairs through the Minister for Emergency Management, the Coordinator-General and the National Emergency Operations Centre (NEOC) for domestic operations and to Foreign Affairs through the Minister for Foreign Affairs and Director General, ASIS, for international operations. Put another way, the CNS is responsible for generic RF preparation, Home Affairs for specific RF preparation and employment for domestic operations and Foreign Affairs for specific RF preparation and employment for international operations.

Diagram 1

Command and Control of the Response Force

National Security Committee of Cabinet

Strategic Level Command

Department of Prime Minister and Cabinet

Chief of National Security

Operational and Tactical Command

Home Affairs

Emergency Management

Operational and Tactical Command

Foreign Affairs

Australian Secret Intelligence Service

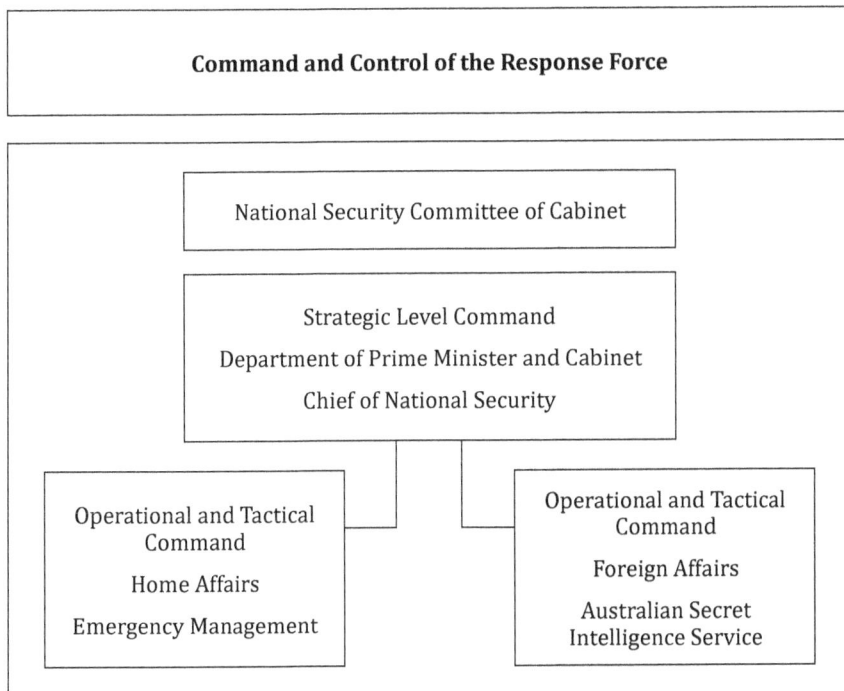

Australia has human capital ranked amongst the highest in the world. Australians are world-ranked scientists, engineers, technicians, corporate leaders, entrepreneurs, lawyers, doctors, actors and sportspeople. Governments should marshal this strategic advantage urgently and astutely to enhance national security for the grey zone. The government can counter an escalating grey zone campaign by identifying and rehearsing specialists in the corporate and public sectors. A fully trained and resourced full-time RF, deployable anywhere and anytime, and a part-time RF working their day jobs and maintaining close community connections optimises Australian human capital for internal security, resilience and homeland defence.

The way ahead

For the book's purposes, let's assume the Prime Minister and the government meet the implementation challenges, remove the Army Reserve, its annual budget outlay, and infrastructure from Defence, and reallocate them to PM&C for employment in the Home and Foreign Affairs portfolios. This consolidation of homeland defence accords with the origins and traditions of Australian part-time military service for territorial defence and the 21st-century enhancement to include the cyber, information and space domains.

Thousands of Australians will rally to the RFR and RF because they prefer to spend their time and talent preventing war rather than fighting one. Australia will send the same messages to them as it does to the CCP.

> If you serve in the RF, you will seek to persuade the CCP to trade and not argue. If the CCP persists in illegal behaviour, the RF will inform the CCP that bullying has uncomfortable consequences. Suppose the CCP ignores RF 'stings' and 'jabs' and invitations to negotiate and persists with coercive escalation towards hybrid warfare; you will be part of the force that wrecks those preparations.

Australian governments will require prompt and astute responses to escalating grey zone attacks on the homeland, or like 2020–21, must also respond urgently with additional resources in the aftermath of natural disasters that displace and distress communities. A full-time RF will have an 'anywhere and anytime' de-escalation mission to oppose an escalating CCP grey zone campaign. The RFR will support the RF mission with a talented part-time workforce embedded in the community to support this mission and bolster Australia's emergency responses to natural disasters, civil emergencies, and the aftermath of grey zone attacks.

The RF and RFR will be a total force cohabitating in shared facilities, collaborating on all operations and coordinating seamlessly with government departments, corporations, and community organisations. The Army Reserve 2nd Division should be renamed the 2nd Response Force Division (2RF Division), comprised of response brigades. HQ 2RFD and its brigades would command disaster response and grey zone countermeasures units primarily trained for cyber, information, surveillance, drone and counter-drone and grey zone operations and secondarily for disaster relief.

Low readiness is detrimental to part-time voluntary national service. The need to counter a grey zone campaign is urgent, and natural disasters occur suddenly. The RFR should comprise high-readiness units on standby for sudden natural disasters, civil emergencies, and contemporary grey zone operations. This posture is like the readiness regime for Australia's volunteer Rural Fire, Maritime Rescue and State Emergency Services. Dual orientations to high readiness employment on exceptional RF operations either on location or online and community-based operations in the aftermath of disasters will attract talented Australians to these new, innovative organisations when and where they are most needed.

Australia has a well-developed and consolidated National Emergency Management Agency (NEMA) within Home Affairs. Interestingly, for our argument about the 'core business' priorities and preferences of Defence and the Army, NEMA's predecessor, Emergency Management Australia, was in the Defence portfolio until 2001 when, figuratively, Defence was assessed to be the wrong hammer for the natural disaster and civil emergencies nails after a revealing national audit in 2000 discovered inefficient and ineffective Australian emergency response arrangements.

The National Emergency Operations Centre coordinates and controls disaster response efforts and acts as the command centre for all communications and information relating to response operations within NEMA. The NEOC also liaises with responsible ministries on national response efforts. PM&C will host a full-time RF Rapid Response Command component. The NEOC will host an RFR Rapid Response component to coordinate high-readiness RFR individuals, teams and units who have rehearsed for national resilience contingencies with State and Territory emergency and rural and metropolitan fire services.

The RFR will be comprised of readiness tiers. The highest readiness tier (First Responders) would be for individuals, teams and units tasked with short notice contingencies, that is, rapid response to sudden events; a medium readiness tier (Second Responders) for anticipated routine contingencies and a lower readiness tier (Third Responders), focused on basic community-based humanitarian emergency operations and supporting events of public significance. These tiers are service options for real and ready capabilities that rehearse annually for authentic contingencies for the nation's defence, security and community well-being.

Contemporary grey zone campaigns integrate political, economic, cyber, electronic and informational actions. Australia's responses must integrate seamlessly to counter these actions. Therefore, RFR First Responders are not citizens with the highest attendance or training levels in the military trade. They are citizens with skills for immediate employment or employment after specific preparation for domestic and international extraordinary grey zone operations. Their skills are maintained and honed in civilian employment to quickly adapt to or be enhanced to counter an escalating grey zone campaign, especially to defend in the cyber and information domains. Remember the Estonian response to their electronic Pearl Harbour in Chapter 2.

Suppose grey zone coercion escalates to destructive cyber-attacks and the deployment of grey zone combatants with capabilities to disrupt Australia's political system, essential services, social cohesion, and the economy and its supply chains. In that case, RFR First, Second and Third Responders are ready for their missions in Australian communities nationwide. This consolidated response is the essence of 21st-century part-time national service.

The RFR is more sustainable and viable if deeply rooted in communities. One root is the number of people serving in the RFR per capita in communities. The other root is protecting communities under stress and displaced by natural or grey zone disasters. For the first root, enhanced and flexible modes of part-time national service will enable the inclusion of more Australians for national resilience within communities.

Multiple modes of RFR service would target underrepresented portions of the population, attract citizens from a broader range of ethnicities and cultures and include persons not traditionally considered to have attributes for military service but whose skill sets justify inclusion for national resilience, especially in the grey zone where sophisticated ICT, including hacking, and information skills are required.

New modes of service would remove access barriers through

broader selection categories, shorter initial training periods, adaptable trade categorisations, flexible medical/fitness requirements, and flexible team structures. The RFR would put civilian specialists into uniform by engaging industry leaders, resident non-citizens, and other groups who aspire to defend Australia but do not currently have the opportunity or pathway.

A 'serve from anywhere' model would leverage online ICT and be designed to avoid anchoring the RFR in the existing depot infrastructure. It targets regions with no RFR service options. Ultimately, more RF Reservists will serve in more communities and workplaces nationwide. This mode simultaneously centralises, diversifies, and disperses facilities while optimising online connectivity to provide a service option from wherever RF Reservists live, whenever they are available and within their means.

Finally, and by no means the least important, the RFR will marshal and operationalise the best human capital within the 1.2 million Australians of Chinese heritage. This community, which does not participate proportionally to its numbers in national security, will play a crucial role in our national defence. The CCP's attempts to influence some members will be countered, and many more members will accept an invitation to put Australia first by becoming RF and RFR members. Their selection, training, and employment will focus on their cultural and ethical preferences, ensuring a strong defence while preventing conflict with China.

Conclusion

Australia's first grey zone battle will not be fought to counter the CCP campaign. It will be fought between those arguing for the NDS2024 as the only option for Australia's defence and those who recognise the grey zone threat and support a Let's Trade, not Argue strategic response to complement the NDS2024. This book makes a compelling case. Those who oppose it and the diversion of resources to establish the RF and RFR need to justify their opposition. We not only welcome but also encourage a robust debate and a better grey zone plan. We oppose appeasement, an NDS2024-only option, or She'll be right, mate' approach that echoes Australia's lack of vision in the 1930s and leaves Australia vulnerable in the grey zone battlespace.

The RF can fit into Australia's national security arrangements with enough public and corporate support and political commitment. Australia has responded effectively when threatened in the past. Perceptions of terrorist and people smuggling threats created the political consensus for establishing the Home Affairs portfolio. Perceptions of the COVID-19 threat prompted a nationwide adaptation and a diversion of resources into countering a pandemic that appeared at the time to be capable of killing thousands of Australians. Responses to natural disasters that threatened life and property also prompted a diversion of resources to protect and support community recovery.

CHAPTER 8

What will the Response Force Do?

Introduction

The Australian Government will not implement a Let's Trade, not Argue strategy unless it knows what the RF will do. All Australians must understand what will be done in their name to keep them safe and what the RF needs to accomplish this mission. The RF's proposed extraordinary operations will be politically and ethically challenging for a liberal democracy like Australia. Consequently, what the RF does must be legal, ethical and morally defensible. Defence and the Army will argue against the Let's Trade, not Argue strategy and resist transferring the Reserves to the RF until they know what the RF and RFR will do that Defence cannot or will not do. Australians will not deter the CCP from escalating illegal behaviour in the grey zone until they comprehend what the RF can do.

Let's begin by taking the term 'special' off the table. The RF is not a revamped special forces organisation designed to conduct traditional special operations. It is not an Australian CIA or a Mossad or MI5 and MI6. We also prefer to separate the notion of 'special' from the idea that only ADF special forces are 'special' and able to conduct 'special' operations. 'Special' is a word that confuses and invites criticism rather than clarifies, specifies and invites admiration. The term 'special' has historical and ethical 'baggage' that we do not want to conflate 'special operations' with ethical, lawful and, most often, non-violent RF extraordinary operations.

It is better to define what the RF will do in a way that broadens the discussion of always connecting special operations with special forces or covert and clandestine intelligence operations with CIA-like organisations. The RF is 'like' special forces in many respects but not 'of 'special forces, and it is not seeking to emulate existing special forces and CIA-like organisations. This book is not copying or upgrading anything. We are applying the lessons of Australian Pacific War history and suggesting unique, extraordinary capabilities for Australia's national security. The book can only offer an extraordinary operations framework for security reasons rather than describe specific contingency operations. The RF game plan must be secret to achieve its objectives. Australians must trust their prime ministers and chiefs of national security and the collective decisions of the NSC to always be in Australia's best interests.

Timing will be crucial. There will be no impetus to build the RF will until most Australians and their political and civil society elites believe the current light-grey CCP political campaign is escalating ominously. The unfolding events the book predicts for 2025/26 should trigger a national conversation about what to do in response. Hopefully, violent, disruptive and shocking events will prompt a call for countermeasures. This 'teachable' moment will occur when Australia's ICT systems go 'on the blink', and the CCP disrupts food, fuel and energy supply chains and severely embargos Australian exports. Australians will have another 1942 moment even though no one is going to bomb Darwin.

Command arrangements

The RF is a fourth armed service that defends Australia in the grey zone. A Hollywood analogy is that it is Australia's 'Mission Impossible' capability that the PM and PM&C have the authority and responsibility to raise, train and sustain. It is not separate and secret from government departments and agencies. It partners with Defence, Foreign Affairs, Home Affairs, and Attorney General's for national security. It will collaborate to achieve the NSC's required specific effects.

There will be three tiers of command: strategic, operational and tactical. The CNS, reporting to the PM, will strategically raise, command, train, and rehearse the RF. For international operations, Foreign Affairs will command and employ assigned RF teams operationally and ASIS tactically. Home Affairs will command operations and tactically employ assigned RF teams with ASIO, the AFP, and the ABF partners for homeland operations. These departments and agencies will host the RF in their infrastructure and facilities, supporting and maintaining assigned RF capabilities when cost-effective.

Phase 1 De-escalation

Let's discuss how to nip Phase 1 bud to remove the CCP's temptations to escalate from 'light grey' political and economic intimidation to darker grey Phase 2 coercion. This description of what the RF will do in Phase 1 sets the scene for how it will de-escalate Phase 2 and 3 escalations if Phase 1 de-escalation fails. What must the RF do to achieve this effect – to disappoint the dragon?

The RF will conduct extraordinary operations by consummate professionals to achieve specific effects. We are not relabelling 'special' as 'extraordinary' to avoid controversy or complexity. A particular response power mindset guides the design and execution of extraordinary operations. This mindset focuses on influencing CCP thinking, not 'blowing stuff up'. These operations have a high informational and psychological impact and persuade the CCP to discontinue illegal activities against Australia's national interests and return to fair trading relations. Extraordinary operations are war preventers, not war provokers and will never be war winners. This response power mindset borrows from the Chinese strategist Sun Tzu rather than Germany's Clausewitz or America's William McRaven. (Google McRaven the Theory of Special Operations)

Detect and stare

Intelligence leads RF operations. The RF must have a close relationship with the NIC to decide what it needs to do, acknowledging that the PM and the NSC authorise all RF operations. The detection pillar is ready, and it is vital. The NIC has evolved into a federation of intelligence agencies, each reporting to the Office of National Intelligence (ONI) in PM&C. Andrew Shearer, the 2025 Director General of Intelligence, will work for the CNS, the RF's commander. Shearer stated at the 2024 Raisina Dialogue in New Delhi, 'grey zone conflict is ... as old as warfare itself ... [The NIC] has a profound responsibility to identify risks before conflicts escalate into open hostilities, particularly during a period where Australia no longer had a 10-year strategic warning window for major conflict.' Shearer can warn about threats but cannot do anything about them. The CNS and the RF can and will do something about them.

The first mission of the NIC and the RF is to detect and stare at threats. The capability to 'stare' secretly at detected threats is essential. The aim is to get Australian surveillance 'in front' of emerging threats. This attribute requires secret planning, preparation, deployment, employment and re-deployment of RF teams and agents. Undercover surveillance and reconnaissance enable early warning in concert with Foreign Affairs and its overseas information-gathering networks and allied intelligence agencies that inform the Australian Government.

RF surveillance teams and agents moving seamlessly into and out of Australia's areas of strategic interest, working in Australia online, and employing 'eyes in the sky' drones and satellites deliver two dividends. The first dividend is a persistent presence overseas that is extremely sensitive to change and variation in emerging and existing threats to Australian national interests. This permanent secret presence would inform the NIC and complement allied intelligence sources. The forward deployment of RF teams to remain in place on rotation in both contested and uncontested environments and governed and ungoverned spaces is a strategic early warning system. This physical, online and 'eyes in the sky' presence is a human and technology-generated 'radar' for political and hybrid warfare threats, especially involving WMD. In Chapter 5, the book quoted an expert who assessed Australia's pre-war and Pacific War coast watchers as an impressive and successful human intelligence network. This history of Australia having eyes and ears throughout its regional neighbourhood in the 20th Century should be repeated in the 21st Century.

The second dividend is the enhanced capability for forward-based RF teams to provide a 'warm start' and immediate action in response to sudden changes in strategic circumstances or unforeseen threats. This rapid response has two specific effects. The first is to de-escalate an emerging security crisis quickly and decisively before it becomes dangerous. This 'nip in the bud' capability can have a deterrent or a coercive effect. Suppose an initial warning and possibly an informational or cyber sting fails. In that case, personnel and resources are in place to execute a pre-emptive 'jab' across all domains – a firmer message that 'enough is enough'.

Secret strategic reconnaissance creates 'escalation dominance' in the grey zone. In effect, the government retains the initiative in the face of a darker grey threat, usually vigorously denied and secretly pursued 'behind the mask', or other dangers and political sensitivities constrain conventional or special forces employment. It maintains the initiative and de-escalates well before there is talk of war or risk of miscalculation.

With de-escalation in mind, the RF must be capable of conducting surveillance and intelligence operations with an 'always on' information collection and targeting mindset. These attributes give Australia the political, diplomatic, military and informational advantage necessary to underpin its competitive, forward deterrence approach in the grey zone. No longer will Australia wait for the CCP's next move. The RF will detect, stare at and pre-empt the next move.

Enhancing the National Intelligence Community

The lessons from the Pacific War apply to the NIC in the 2020s. Australia must bolster human (HUMINT), signals (SIGINT), and imagery (IMINT) intelligence capabilities immediately to detect and stare. The Australian government needs to know where the CCP's grey zone workforce live and work and how to message them when and where Australia wants to do so. Australia must know where China's armed forces are and how home bases work. That knowledge and the ability to target people, places and things to de-escalate in Phases 2 and 3 are Australia's ultimate deterrence. What are we talking about for detecting and 'staring' – surveillance?

The NIC must combine HUMINT, SIGINT, and IMINT to identify human, electronic, and physical targets to inform RF operations. HUMINT is its ability to obtain intelligence directly from and about people, including their intentions, plans, and attitudes, and to network sources for broad coverage. SIGINT draws intelligence from or about various forms of communication through intercept and decryption of written, oral and electronic communications using space, aerial, maritime or land-based intercept systems. Experts and Artificial Intelligence typically analyse networks and gather information about what key individuals are communicating and what command and control systems are doing.

SIGINT now includes computer network exploitation (CNE), i.e.

penetrating ICT networks and databases to extract intelligence. SIGINT's access also creates the opportunity for Computer Network Operations (CNO) and Computer Network Attack (CNA), as discussed below. IMINT analyses imagery gained by space, aerial, maritime and land-based platforms. Typically, SIGINT and IMINT discover the location of conventional, unconventional and grey zone combatants, and HUMINT and SIGINT discover what they are thinking and doing.

Bring on Australia's 21st Century coast watchers and code breakers! They won't be individuals with binoculars, a radio, Morse Code, or boffins applying complex math formulae to crack codes. The NIC and RF workforce will not only be super-clever spies, linguists, cultural experts and analysts. They will include super-clever geeks, nerds, hackers, spin doctors, pollsters, and marketing gurus. They will combine humans and technology to see and hear everything in Australia's area of strategic interest. Australia must know what is happening in its regional backyard and not rely on allies to tell us what's happening. Fortunately, the Americans, with shared interests in understanding what is going on in China and the Asia-Pacific region, have no reason for not sharing their intelligence with us if we remain a loyal ally. The AUKUS Agreement facilitates Australian access to US and UK intelligence. Australia needs to know how to use allied intelligence wisely. The RF/NIC collaboration is an additional sovereign capability that will keep Australia safe.

Informational actions

Informational actions will cue and contribute to the 'sting' of RF operations. The Government can take informational actions once the NIC detects and the RF 'stares' at the CCP campaign. The Home Affairs and Foreign Affairs portfolios are Response Power's informational pillars. The 'battle of the narratives' is about understanding, shaping, and influencing public opinion and the decisions of governments. Australia is smart enough to shape international public opinion and perceptions: revelations and information directed at the CCP campaign 'hurt' by discrediting biased CCP narratives and exposing illegality.

The Phase 1 messages are, 'We know what you are doing, and we are telling the world about your illegal activities. Please stop, and then we will have nothing to say to make you uncomfortable.' Diplomacy and negotiations about how to 'get along better' accompany informational actions that reveal illegality. Detection and surveillance do not counter the CCP grey zone campaign until there is astute media exposure of CCP activities. The simple and persistent message is, 'Let's trade, not argue.'

Public exposure of the CCP campaign delivers low-level deterrence in Phase 1 but may not be enough to entice the CCP to concentrate on trade, not escalating interference and control. The more significant challenge is to deliver increasing levels of deterrence with uncomfortable RF 'stings and jabs' in response to darker grey escalation. Suppose diplomacy and negotiation fail and CCP grey zone escalation continues. In that case, Australia can only deter and de-escalate if the RF is 'ready, willing and able' to sting and jab to get attention. The aim is to persuade the CCP to stop illegal activities, such as the cyber infiltration of and attacks on Australia's ICT systems – the invisible invasion – and espionage, subversion, political interference and secret police activity on Australian soil – the silent invasion.

Informational actions supporting RF operations can be proactive and reactive. The NIC allows Australia to say proactively to the CCP, 'We know where you are and what you are doing'. The RF allows firmer messaging, 'If you do not stop illegal activities, there will be uncomfortable consequences, so let's discuss how to get along and trade, not argue.' The RF facilitates the message after it acts, 'Look what you made us do in response to your persistent illegal actions. Please stop bullying. We don't want to be unfriendly again. Let's discuss how to get along. Let's trade, not argue.'

Attack or just defence?

Should Australia conduct offensive informational actions or leave the information warfare initiative with the CCP? Defensive actions to protect information are accessible and lawful. Australians are comfortable with active measures such as information and personal security, data security, facilities security, cyber security, and non-violent counterintelligence operations.

Offensive informational actions counter threats and aim to change the minds of CCP hardline nationalists. The aim is to influence decision-making with informational actions. These actions include public and covert information, propaganda and disinformation to shape CCP emotions, attitudes, motives, thinking processes and behaviour. The primary emotional target is to create fear in the minds of CCP leaders and grey zone combatants and operatives. They must know and feel the consequences of their illegal behaviour or leading and manipulating others to behave illegally against Australia's national interests.

We recommend 'virtue signalling' for Phase 1 de-escalation. Australia is proud of its heritage, values and enterprise. Its combined Indigenous, settler and immigrant Australian identity is characterised by a fair go and having a go. Though Australia has a significant military history, we are not an aggressive people seeking to dominate others. We want to get along with everyone but will defend ourselves if anyone intimidates us or our regional neighbours. Australia does not force democratic values on others. The message to the CCP is, 'Please respect our political differences, sovereignty and alliance choices, and we will respect yours. Let's trade, not argue.' It is absurd for Australia to lecture China on its government's conduct. The Australian Government would be just as irritated if China did the same.

Phase 2 messages

We recommend 'fair warning' messages for Phase 2 de-escalation. This escalation would see the CCP continuing espionage and political interference and escalating to coercion through economic embargos, increased numbers and severity of cyber-attacks and physical and social media harassment of individuals and groups. Messages during Phase 2 de-escalation operations would be.

> We warned you that escalating your illegal activities to violence and destruction would have consequences, and you didn't stop. You ignored our invitation to trade and not argue, forcing us to demonstrate uncomfortable consequences. Please accept our invitation to trade, not argue, or there will be more discomfort.

Phase 3 messages

We recommend a 'regrettable firm actions' message for Phase 3 de-escalation. This escalation would see the CCP deploying armed operatives to Australia to stir up trouble, either directly through violent action or, more likely, by coercing, manipulating or persuading others to act violently to unsettle Australian society under the guise of opposing anti-Chinese groups and promoting groups campaigning for peace and prosperity between Australia and China.

Cyber-attacks would increase in frequency and severity. These attacks disrupt supply chains and essential services to the point where power outages and a lack of food and services distress vulnerable groups and displace some communities. Trade embargos, harassment of Australian ships and aircraft, and the ominous appearance of Chinese naval vessels and service aircraft near Australia's trade routes cause a stock market crash. This crash verifies Phase 3 economic damage that has cut government tax revenues and put thousands of Australians out of work.

Australia's messages during Phase 3 de-escalation operations would be:

> You have been planning to be violent all along. We are telling the world what you are doing. Once again, you have ignored our invitation to stop your preparations for violence and disruption and to trade and not argue. You forced us to show you the consequences of planning violence and to attack our economy and way of life. We have been firm in causing discomfort because you have misbehaved towards us. Stop your hostilities towards us. Let's trade, not argue.

Extraordinary operations

Extraordinary operations are active measures to achieve or support political or national security objectives. The RF will conduct extraordinary operations by consummate professionals anywhere and anytime to achieve specific effects. It does not project physical force and firepower like the ADF to take military action. It is about projecting small teams and agents, secretly and disguised, to designated locations in Australia and overseas and then across sea and land, upwards into the air or below the water surface for extraordinary operations. Other teams and agents operating online and through telecommunications won't go to designated locations. Still, they may assemble face-to-face at headquarters for coordinated action to achieve a specific defensive or offensive effect on the Internet, in the media or ICT systems.

One of the most famous historical examples of an extraordinary operation by consummate professionals to achieve a specific effect was using a wooden horse to infiltrate a small team of warriors into the ancient city of Troy. Conventional operations had failed. This extraordinary operation, conducted by the bravest – arguably 'crazy brave' – and most capable warriors, achieved an amazing specific effect – immediate uncontested access to Troy after years of siege, suffering and inconclusive bloodshed. This operation followed Sun Tzu's advice to understand the minds of one's enemies as well as their military dispositions. The Greeks expected the Trojans to take the wooden horse into the city as a trophy. They may have 'leaked' the idea – an informational action – that it was a Greek acknowledgement of defeat to be prized and put on display, not burnt.

Another example that did not involve special forces conducting special operations was the 'Dam Buster' raids against Germany's Ruhr Valley dams in the Second World War. A few aircraft with new technology did not cause significant physical damage to the targeted dams. Still, the negative psychological impact on Germans, the positive psychological effect on the Allies and the diversion of German resources to protect repaired dams were substantial. The Doolittle Raid on Tokyo by a few US aircraft in April 1942 did minimal damage to Tokyo but had significant psychological and strategic repercussions. Both were Powerful Owl Phase 3 extraordinary operations because their psychological dividends were at least ten times their investment. Neither the Dam Buster nor the Doolittle raids could be repeated for Phase 3 de-escalation in the 21st Century because they were acts of war after the war was declared. Their 21st Century Phase 3 equivalents could be well-publicised red kangaroos stencilled on sensitive facilities in China and swarms of drones dropping brightly coloured pamphlets with pictures of grey zone combatants and stories of CCP grey zone activities over government buildings in Beijing coinciding with the same information being disseminated in the media, on social media and YouTube for a worldwide audience.

Ultimately, extraordinary operations become lethal after the CCP has caused the deaths of Australians either directly through assassination or terrorism or through destruction and disruption of supply chains and essential services, such as power, fuel and medical services. An American example is the de-escalatory effect of killing Qasem Soleimani, Commander of Islamic Revolutionary Guard Corps Quds Force, at Bagdad airport on 3 January 2020 using an MQ-9 Reaper drone on the orders of President Donald Trump in response to an accumulation of Iranian provocations and Soleimani's Qud Force responsibility for killing hundreds of American and allied troops in Iraq. A US Special Forces team followed Soleimani's car and, after it was hit, examined the vehicle, took photographs and video footage of his body, torn limb from limb, and his possessions, and then disappeared. The White House released an aerial video of the strike and the strike team's photographs and footage to Fox News.

The United States broadcast Soleimani's demise to the world. The deterrent message to leaders in Iran appeared to be, 'We can find you and eliminate each of you if you continue to provoke us.' An estimated one million Iranians turned up to Soleimani's funeral to voice their outrage. The Iranian government had to do something. It retaliated with rocket attacks from Iranian soil onto bases in Iraq containing US personnel on 8 January that did not appear to be intended to cause substantial American casualties but did damage some infrastructure. Iran's deterrent response to the United States seemed to be, 'We are angry, and our people need retaliation, but we are not giving you sufficient retaliation to escalate to conflict'.

After these incidents, there was a brief war of words in the media, but no further escalation of hostilities. Iranian special forces stopped harassing Saudi Arabian oil fields and attacking oil industry tankers. – the 24-hour media cycle moved on. (Google Assassination of Qasem Soleimani) Neither the United States nor Iran were at war, and neither were declaring war. They were employing grey zone tactics to achieve political ends. President Trump used an exceptional strike operation to de-escalate tensions for a specific strategic effect.

Pacing

Exceptional operations maintain the initiative by progressing through the three expected escalation phases of a CCP grey zone campaign. The goal is carefully calibrating these phases to outpace CCP escalations and deter further actions. Australia must not allow the CCP to strike first repeatedly.

Pacing begins with detection and observation. Understanding what the CCP is planning and preparing helps to outpace any malicious taking campaign. Exceptional operations should be calibrated to deter escalation before rather than responding after an attack. The most effective message to convey beforehand is, 'Please don't be unpleasant,' rather than reacting with, 'You have forced us to be unpleasant because you have been unpleasant.'

Covert and clandestine

Extraordinary operations are usually covert (hide the sponsor's identity or plausibly deny sponsorship) and often clandestine (conducted secretly to hide the fact that the activity is taking or has taken place). Covert and clandestine influence operations (activities that use information, psychological methods and relationships to change minds in China) are non-violent. Direct and indirect covert and clandestine operations can become lethal if CCP escalations are about to or have already incorporated violence and destruction. Bullies who hurt must be hurt in return, but selectively, proportionally and repeatedly to alert but not alarm, to force second thinking, and not prompt disproportional retaliation.

Phase 1 de-escalation is about covert influence. These extraordinary operations involve psychological methods to persuade the CCP to end its grey zone campaign and return to conventional diplomatic and trading relations. The aim is to shape emotions, attitudes, motives, thinking processes and behaviour. These operations change perceptions so that the CCP's decision-making and actions change from coercion to cooperation. They also alter international perceptions to attract support for Australian sovereignty because it is the CCP seeking control illegally, not Australia. Others should see that what Australia does with its RF is a 'fair thing' in response to illegal behaviour.

HUMINT, SIGINT and IMINT go hand in hand with covet and clandestine actions. China and Australia are both entitled to their sovereignty. The CCP is conducting covert and clandestine influence operations in Australia to change minds in favour of CCP interests. Australia has no intention of persuading or coercing the CCP to act in Australian interests. Still, the RF's ultimate deterrence is to disrupt hybrid warfare preparations at their source. Therefore, RF covert and clandestine influence operations for this defensive purpose are not 'off the table'. The RF must be capable of extraordinary operations anywhere and anytime to defend Australian sovereignty and send a message that escalations against Australia will hurt. Enough said for now for obvious security reasons.

Cyber warfare is at the core of grey zone campaigns because it can devastate ICT-dependent economies and distress and displace populations. There is no need for physical invasion. The question for Australia is whether to defend against cyber-attacks exclusively or use offensive cyberattacks as a deterrent. The CCP rehearses for electronic Pearl Harbours with 24/7 cyberattacks and infiltrating ICT systems for timed activation to disable them. Should Australia deter this behaviour by stinging back in the cyber domain?

Offensive cyber operations

The argument regarding RF covert and clandestine influence operations also applies to cyber operations, encompassing computer network attacks (CNA) and exploitation (CNE). CNE serves as a form of non-violent action because it does not disrupt or destroy systems. In contrast, CNA represents violent action, as it aims to penetrate computer systems to destroy or disrupt critical operations, deny access to data, or corrupt, alter, or falsify information. CNA specifically targets government or military critical information, command and control networks, control systems, and the functioning of selected public or private infrastructure and other production facilities.

Specific infrastructure and production facilities could include ports, airports, power supply, transport, media publishing or broadcasting facilities, or selected manufacturing facilities, especially where they produce WMD components or other weapon systems. Australia must acquire concealable and mobile high-energy radio frequency weapons for CNA and employ electromagnetic pulse technology. Enough said about these RF capabilities for now.

More broadly, CNOs suit the RF's extraordinary operations because they are cheap and send messages that disrupt rather than destroy. They ignore geography and access targets anywhere and anytime. It is not suggested that the RF will develop parallel SIGINT infrastructure and capability to ASD. Instead, Australia must have unapparelled coordination in the cyber and online dimensions to ensure integration and deconfliction to achieve multiplier effects.

The ethical downside is that disrupting ICT systems can cut energy, food and water supply and cause death, distress and displacement. The CCP may have no qualms about this tactic that emulates Russia's cyber warfare. Australia should have qualms and use these cyber technologies only as a last resort if other stings and jabs do not create sufficient deterrence and invitations to talk and trade are ignored. The message is that Australia protects its sovereignty and wishes to return to a trade-based relationship. There is no ill will against the Chinese people and their well-being, but Australia must hold an olive branch in one hand and a 'taser' effect in the other.

It is all about messages. If initial stings and jabs fail to stop the CCP escalation of illegal actions and preparations for hybrid warfare are discovered, the message is,

> We warned you to stop bullying, and we have now discovered that you want to get violent. You are forcing us to interfere with your preparations, but we are always ready to discuss how to get along. However, it would be best if you stopped preparing for hybrid warfare.

The RF must be ready to disrupt hybrid warfare preparations. The final message is,

> We know where you are and are coming for you because you have decided to go for us.

Direct and indirect approach

Two features of extraordinary operations are a direct and indirect approach. The direct approach focuses on Australian activities to change minds in China. The indirect attribute focuses on persuading others to change minds in China with their own extraordinary operations. In short, the direct approach involves our stings and jabs, and the indirect approach involves enabling the stings and jabs of others.

Direct effect operations are the precise execution of activities anywhere and at any time. They employ small teams in hostile, denied, or politically sensitive environments to influence and de-escalate threats in the first instance and seize, destroy, capture, exploit, recover, or damage designated targets if warnings are ignored and escalation continues.

Indirect enabling operations are activities that involve a combination of lethal and non-lethal actions taken by specially trained and educated teams that have a deep understanding of cultures and foreign languages, proficiency in direct effect de-escalation tactics, and the ability to operate with and alongside allied security agencies and other third parties in permissive, uncertain, or hostile environments if warnings are ignored and escalation continues.

Targeting

After detecting, staring and warning, precision targeting is the essence of RF operations if warnings are ignored. Targeting is the stern Phase 3 end of Australia's Response Power when deterrence is insufficient to stop escalating coercive behaviour. One attribute is an ultimate sanction to 'hurt' that hardens deterrence. The other denotes enabling operations with allies to empower them to establish credible deterrence through having their own ultimate sanctions that 'hurt'.

Aligned with pacing, the CCP must 'feel' an RF operation psychologically rather than just knowing there are uncomfortable consequences. An example mentioned earlier is red kangaroo stencils appearing on sensitive facilities and during public events as a warning. This extraordinary 'graffiti' operation is non-violent and may seem frivolous. Still, it creates the fear of what might happen if those who organised the stencils decided to use their access to a facility or public event to dispute it violently. That's an example of a messaging 'sting' to support deterrent diplomacy.

Consummate professionals

We are not relabelling special forces with a new elite title. The RF's consummate professionals are not super-fit Anglo-Celtic daredevils with specific military skills and a particular special forces culture or 007 agents. Instead, they are consummate professionals in all walks of life, some operating in high personal risk environments and others at no personal risk. At high and medium risk are the 'best of the best' bomb disposal and explosives technicians, firefighters, emergency rescue teams and police special weapons operations squads. At minimal risk are IT technicians, marketing executives and intelligence analysts. All have individual competencies and attributes to use technologically advanced processes and equipment and receive sophisticated training to enhance their complex problem-solving attributes. All will have gained membership based on merit and attributes for extraordinary service. There will be an emphasis on reaching out to Australians of Chinese heritage to deter their mother country from coercing their country of birth or choice. Specific skills and capabilities resident in Australia's special forces and elsewhere in Defence may be leveraged to deliver particular effects within Phase 3 RF operations when appropriate.

While the exact reasons and selection processes for the warriors who undertook the extraordinary operations, such as the fall of Troy, the Dam Busters, or the Doolittle air crews, may never be fully understood, we can conclude that these individuals were well-adjusted, mentally and physically fit, intelligent, well-trained, well-educated, and experienced. They were chosen to minimise uncertainty and risk while maximising the chances of success.

The RF's challenge lies in developing effective selection, induction, training, rehearsal, and management systems. These systems are essential for discreetly recruiting talented Australians and transforming them into consummate professionals. They involve comprehensive curriculum design, appropriate facilities, secure locations, skilled instructors, dedicated mentors, and advanced technology.

Three key concepts – Ghosts, Guardians, and Goblins – represent the RF's top professionals in their efforts to achieve specific objectives.

Ghosts symbolise the direct-action intelligence, informational, and precision targeting aspects of RF operations. They utilise advanced 'ghost' technologies to carry out extraordinary missions across all domains.

Guardians embody an indirect approach to encouraging, guiding, and empowering allies in Australia's nearby region. Their focus is on developing de-escalation strategies and response forces that align with the allies' preferences for countering the CCP's bullying and manipulation.

Goblins are highly skilled professionals who specialise in conducting exceptional operations in the fields of information and communication technology (ICT), information, cyber, and space. Goblins and Ghosts often collaborate.

The RF's ghosts, guardians and goblins will be responsible for creating a range of specific effects or, metaphorically, stings, jabs and pokes that pace CCP escalations but can become harder punches for Phase 3 de-escalation.

The Ghosts – Covert and Clandestine Operations

Covert and Clandestine Operations [CCO] are RF activities for defending against and operating in the grey zone when no war is declared, but hybrid warfare is in the offing. CCO focus on secret forms of small-group penetration anywhere worldwide for as long as it is required to gain situational awareness and target information. CCO can occur both in the physical (maritime, land, air) and through cyber and space domains. They must be undetectable, moving in and out of areas without notice. They must be 'ghosts' who are mysterious fear creators and never truly discoverable.

The ghosts operate in the fog of the grey zone to defend against the sponsors of unethical and duplicitous hybrid warfare and other operations. Hybrid warfare is persistent and population-centric, emphasising espionage, sabotage, and intimidation, combined with covert action, information warfare, and, when required, proxies, partners and influencers. Countering this mix of illegal activities is the RF CCO realm.

This permanent presence in sensitive places worldwide means Australia will have what it does not have now: a 'first mover' option. The CCO consummate professionals are experts in deterrent statecraft across a spectrum: competing for influence, creating psychological effects on one end, and applying force for strategic impact at the other.

The Guardians – Neighbourhood Watch

The CCP's grey zone campaign targets Australia's neighbours. The Guardians conduct enabling operations involving collaboration with these neighbours and aggregating their contributions towards a regional de-escalation strategy. These enabling operations help others to empower themselves to oppose the CCP's grey zone campaign.

This strategy relies on a long-term plan to shape, understand, and influence Australia's near region and other areas of strategic interest. This indirect approach aligns with the spirit of NDS2024.

Enabling operations involve clandestine personnel working in Australia's backyard. This long-term engagement campaign focuses on identifying and investing in people and infrastructure to shape and influence political elites, civilian agencies, regional military forces, and law enforcement agencies. The enabling "tradecraft" includes mentoring, providing training assistance, enabling the use of niche technologies, targeted diplomatic influence, and capacity building through specialised aid.

These personnel contribute to joint, interagency, and alliance teams to destabilise a developing Phase 3 CCP hybrid campaign. This destabilisation serves to deter escalation to hybrid operations. The resulting 'escalation dominance' establishes a position of strength against the CCP's unconventional political and information warfare efforts. This approach acts as a deterrent, allowing other elements of Australia's national power to take the initiative and expand their own competitive space.

The Goblins – Electro-Magnetic Spectrum Operations

Ghosts and Guardians alone are not sufficient. The emerging dynamics in the space, cyber, and human information domains will play a crucial role in determining the success or failure of grey zone campaigns. Australia needs highly skilled professionals in these domains to effectively conduct offensive and defensive electromagnetic operations (EMSO) and information operations globally. To achieve high-impact effects on the CCP's extensive, conventional military and unconventional special operations capabilities, RF teams must be able to operate seamlessly across these domains. For instance, covertly penetrating information and communication technology (ICT) systems is one tactic.

EMSO enhances Australia's resilience against the CCP's 'information confrontation capabilities' by mapping cyberspace to conduct deniable cyber offensive operations (CNO, CNE, and CNA) as required. CCO and enabling operations, including EMSO, address highly contested threat areas involving cyber disruption, hostage rescue, high-value, time-sensitive targeting, and setting favourable conditions for renewed negotiations.

EMSO adds a crucial dimension to Australia's independent reconnaissance and intervention capabilities. The technological elements in the grey zone are expected to evolve, including weapons developed based on new information-age principles. One aspect will be Australia's own 'information confrontation capabilities', which include the use of 'deepfake' propaganda. This type of narrative could influence the CCP's will, emotions, behaviour, psychology, and morale.

Artificial Intelligence is a critical component of EMSO. RF teams can deploy swarms of self-learning, autonomous machines to manage sensing, movement, targeting, and communications away from vulnerable areas and extend their reach across vast, dispersed networks. As a result, the CCP will no longer be able to focus on a few significant targets; they will have to search for many more across larger spaces. These swarms will create tempo, deception, and mass, contributing to countering CCP cyber-attack systems. While complex, RF teams must implement these tactics to keep Australia safe.

Conclusion
There is a saying, often attributed to George Orwell or Winston

Churchill, that 'People sleep safely in their beds because rough men stand ready in the night to visit violence on those who would do them harm.' This book does not focus on these 'rough men' applying lethal force. Instead, the RF combines intelligence, situational awareness, and ingenuity to utilise a diverse range of skilled professionals who conduct extraordinary operations in any domain, at any time and in any location to ensure the safety of Australians.

In 2023, Kim Beazley, the former Defence Minister, unintentionally referenced the RF's Goblins while endorsing John Blaxland and Clare Birgin's book, *Revealing Secrets: An Unofficial History of Australian Signals Intelligence and the Advent of Cyber*, keeping the 'rough men' quote in mind.

> We [Australians] sleep safer because of the 24/7 intelligent, technologically competent, patriotic men and women who work for our agencies, and they develop and work our electronic defence and offence capacities to world-class standards. This is in a world now in which we are constantly under attack.

Beazley, one of the founding fathers of the 1987 'defence of Australia' strategic orthodoxy that reappeared in NDS2024, recognises that the cyber domain changes who and how Australia is defended. The information and space domains do the same. The RF defends Australia in these new domains to avoid war, but like the Powerful Owl, it knows when and where to strike to visit violence on those who would do Australia harm.

Australia's challenge is to capitalise on knowing what is happening in the grey zone and then doing something about it. The NIC warns about the CCP grey zone campaign but cannot respond with sufficient deterrence or de-escalation. The NIC is not a response option. The NIC collects intelligence and feeds the government's assessment and decision-making process; it is not a federation of organisations that can act against grey zone combatants.

The NIC has detected and warned about the CCP's intelligence operations and identified key operatives. It collaborates with law enforcement agencies, typically the AFP, to apply legislation authorising the arrest, detention, trial, and punishment of CCP grey zone combatants behaving illegally. That's low-level deterrence for Phase 1 but insufficient for Phases 2 and 3.

The RF hammer can hit the grey zone nails in Phases 1 and 2, facilitating an astute pre-emptive response in Phase 3. The RF's CCO, supported by EMSO and enabling operations, constitute the essence of what the RF will do. This approach integrates diplomatic, informational, military, economic, financial, intelligence, and legal efforts to shift the focus from responding to CCP attacks to anticipating and preventing them. They complement the ways and means of Australia's diplomatic, informational and economic power and hard military power (maritime, land and air).

Building an RF depends on the political will to direct extraordinary operations to act early, persistently, and lethally as a last resort. On the one hand, RF operations come with political and physical risks. These risks increase when operating with and through international partners and proxy forces. It is essential to know the risks before taking them. On the other hand, Australia's RF will be a risk mitigator. It does not seek or invite a military contest. It mitigates the risk of conflict and slows the bullying momentum.

At worst, compromised covert and clandestine activities will cause some political embarrassment. Still, all RF actions are being taken to ease tensions, de-escalate, and return to a trade-based relationship. There are no sinister warmongering or authoritarian motives to discover. Australia is not seeking control over any nation to establish a tribute state. Australian development aid is donated to facilitate a nation's sovereignty and societal well-being, not control. The rationale for RF operations is to deter bullying and get back to trade.

In summary, an RF enables Australia to identify threats and create appropriate, well-calibrated responses that can achieve deterrence at any location and time. It gives the Australian government more alternatives to conventional or special forces for protecting sovereignty and national interests. The RF actively warns against coercive actions, while the Foreign Affairs Department and Australia's political leaders seek dialogue and negotiations to restore mutually respectful trading relationships.

The history of the lost opportunities for SOA/SRD operations during the Pacific War is instructive. Still, hindsight from 1942–45 is not a strong argument for enhancing Australia's strategic options and developing an RF in the 21st Century. Hindsight is cleverer than foresight. However, we would like to ask some hypothetical questions to conclude this chapter and set the scene for the next chapter.

What if Australia had bolstered the Singapore Strategy by establishing an SOA/SRD as a fourth armed service in the late 1930s and the first two years of the 1940s? What if SOA/SRD had mobilised for Phase 3 de-escalation and was ready in 1942 after the final strategic warning bell tolled on 27 September 1940 when Germany, Italy and Japan signed the Tripartite Pact, formalising the Axis Alliance? Would Japan have had second thoughts about bombing Darwin and invading the Melanesian archipelago if SOA/SRD, after the Pearl Harbour attack in December 1941, had destroyed significant numbers of Japanese ships in harbours and aircraft on airfields in Southeast Asia before and after the fall of Singapore in March 1942?

CHAPTER 9

Next Steps?

Justification

The strategic alarm bells are sounding for the US-led Western alliance about Chinese and Russian strategic ambitions in their chosen spheres of influence. (see Chapter 1) Borrowing from David Kilcullen, the Cold War dragons are back and have learned to fight the West on their terms in the grey zone. Imagine Australian behaviour when ICT systems go down, the lights go off and food, water and petrol supply chains malfunction. Imagine how the nation will divide when a political group emerges favouring better relations with China to secure a peace and prosperity agreement. For people who became unsettled over toilet paper supplies during a pandemic, an escalation of the CCP's political, economic and cyber coercion might provoke anti-social behaviour in Australia that divides communities and brings down governments. (see Chapter 2) The Australian government needs to do more to prepare Australia to meet the CCP in the grey zone. She won't be right, mate. The ADF will not have a navy, army or air force to fight during a silent, invisible invasion. The ADF will watch on while the grey zone campaign escalates politically, economically, and in cyberspace. (see Chapter 3)

No sensible person can deny the CCP grey zone threat. There are extensive ASIO warnings, expert literature and commentary in the media. However, there has yet to be a national conversation on whether the CCP will escalate against Australia again, as it did between 2016 and 2021. No one wants to offend the CCP and bring on trade embargoes again. The book prophesies escalation in 2025/26 but may not be believed like the ancient prophet Cassandra. Fair enough. No one can accurately predict the future. Australians have yet to endure significant political, economic, cyber and social coercion, which is evident in Hong Kong, the Philippines, Taiwan, and the South China Sea.

When escalation is obvious and uncomfortable, this book gives the government and the Australian people a Let's Trade, not Argue strategy to de-escalate. (see Chapter 4) It describes a full-time and part-time Response Force (RF and RFR) framework. The book does not copy any other nation's response to the grey zone threat. It is inspired by lessons from Australia's participation in the Pacific War and a transformed 21st-century voluntary national service scheme that bolsters homeland security. (See Chapters 5 and 6). It is time to toughen up for tough times ahead. Australia needs the RF and RFR to counter the CCP grey zone campaign in Australia and its regional neighbourhood. (Chapters 7 and 8)

Australia must not repeat the wishful thinking of the 1920s and 1930s and hope the CCP will continue its charm offensive after Australia rededicates to the US alliance during Donald Trump's second term. Hope is never a strategy. Still, the government does not have the 'teachable moment' or public pressure to take advantage of the CCP Phase 1 honeymoon in 2025 and embrace Response Power, adopt the book's Let's Trade, not Argue strategy, and urgently establish the RF and RFR in anticipation of the need to deter the CCP from escalating to darker grey Phase 2 intimidation and Phase 3 coercion and disruption. The CCP may continue to court and not coerce in 2025, so the chances of any action are remote. But there is merit in debating and not ignoring this book.

Inconveniently, the time for debate and consensus development is not unlimited. To 'talk the talk' for the remainder of the 2020s and not 'walk the walk' repeats the sleepwalk to war in the 1930s and the haphazard development of the SOA/SRD in 1942 after strategic surprise. The RF needs to be developed urgently as a blueprint or contingency, and planning must be completed to establish the RF in anticipation of the teachable moment. The RF needs to be a contingency plan ASAP, not hurriedly developed from scratch after Australia panics under pressure in the grey zone.

Steps
Prime Ministerial leadership. In many ways, the Prime Minister is also the Minister for National Security and Well-being because

Australians hold them to account when things go wrong. The current or future Prime Minister must champion the Let's Trade, not Argue Strategy and get Australia ready in the grey zone.

Step one is for Prime Minister Anthony Albanese or his successor to bring the National Security Committee of Cabinet together to discuss this book after it has been evaluated in PM&C and other government departments and agencies and debated in the public square. The authors will be honoured to assist. Each department will likely provide contested advice and warn against the book's prescriptions. Defence officials may argue that espionage and foreign interference are Home Affairs and Foreign Affairs problems and something the NIC should address. The Navy and Air Force may argue that developing an RF and RFR is an Army problem. The Army may agree that foreign interference is a Home Affairs issue but argue against an RF, suggesting that Home Affairs counter the grey zone campaign in Australia and Foreign Affairs do so overseas. There will be no appetite for offering ADF special forces to establish the RF. Defence will argue that they should focus only on counter-terrorism at home and counter-insurgency and counter-terrorism operations overseas, as well as responses to sensitive incidents and watching out for WMD.

Home Affairs may argue that an escalation to hybrid warfare is a military threat and, therefore, a Defence problem. Foreign Affairs may support muscling up ASIS to counter foreign interference but shy away from conducting extraordinary operations employing the range of capabilities specified in Chapter 8. Diplomats may also argue that RF capabilities will send too strong a message to the CCP that will provoke further economic retribution and other types of bullying behaviour.

Hopefully, like John Howard in 2003, who 'banged departmental heads together' for the Solomon Islands intervention, Anthony Albanese or his successor will comprehend the motives of departmental opposition, secure Cabinet consensus and take Step 2

Inter-Departmental Task Force. The second step is establishing an Inter-Departmental Task Force [IDTF] involving Federal, State and Territory representatives to produce unclassified and classified versions of Australia's de-escalation strategy and Response Power requirements. This book's authors stand ready to contribute. The government should communicate this plan publicly in a National Security White Paper and secretly in an operational framework and capability development plan.

The departments and agencies securing Australia must come together to provide the ways and means for projecting Response Power and building the RF and RFR. Each instrument of Australia's national power would contribute 'capability bricks'. The Navy and Air Force will assign vessels and particular aircraft to the RF and purchase, operate, and maintain specialised 'ghost' maritime, land and air capabilities. If Defence does not wish to support the RF or can't get things done quickly enough, the government can look to corporate Australia to get the job done.

The IDTF will specify the process for transferring the Reserves from Defence to Home Affairs and transforming the Reserves into the RFR. The first step will be to propose a powerful RFR Task Force comprised of public and private sector leaders and experts to create the RFR.

Legislation. The third implementation step is communicating this new strategy with legislation, laws and the media to shape friends' and foes' thinking about Australia's intentions to defend its sovereignty and national interests using a war prevention strategy. The Australian public should know that its government is strategically astute and proactively developing capabilities to protect them against CCP political warfare in the grey zone by leveraging trading relationships.

The strategic message is 'Let's Trade, not Argue', but Australia has an RF and RFR that will deliver uncomfortable consequences and better resilience if that invitation is ignored. Australia will have legislation and budget allocations that direct an RF and RFR to work with departmental and agency partners to identify threats to Australia's national interests and de-escalate them before they harm – precautionary and ethical but potent.

Rehearsal. The fourth step is to rehearse Response Power, with enhanced force projection capabilities to integrate with and sharpen Australian national power's diplomatic, informational, military, and economic tools. Notably, the RF must have the autonomous capabilities to deploy anywhere and at any time covertly and clandestinely. They should not depend on departments and agencies whose speed and agility cannot meet their requirements. The preparatory functions of Response Power projection must be ready and capable, not the first urgent idea when strategically surprised, as was the case on 6 December 1941 and 11 September 2001. The operational functions and organisational muscle groups must be well-developed and 'stress-tested' with robust contingency rehearsals under pressure to enable effective employment where and when necessary.

Consummate professionals. Concurrent to taking steps 3 and 4, it is time to marshal Australia's human capital for national defence in increasingly uncertain times that echo the 1930s. There is no reason why Australia cannot develop the most sophisticated, technologically enabled detection and deterrence arrangements supported by the most effective RF in the world. Many educated and impressive Australian men and women have the right attributes. Hopefully, Australia's civilian and military leaders have the imagination, commitment, foresight, and ingenuity to meet the challenge of recruiting, training and rehearsing them.

Training system. Finally, Response Power needs a whole-of-government education and training system. This step is about integration rather than separate institutional development. Defence, Home Affairs and Foreign Affairs have workforce education and training systems. These systems are for the core functions of those departments. Accordingly, the IDTF must identify the expertise required for the Response Power workforce and outsource the training needs analysis to lay the foundation for the design, development, delivery and evaluation functions of a Response Power education and training system. The intention is to enhance well-established department and agency education and training systems to skill the RF workforce.

Affordability

An RF is affordable for middle and smaller powers. The ways and means are about human performance with technical augmentation rather than ships, submarines, tanks, artillery, and strike aircraft using human operators. It is about consummate professionals vs. CCP grey zone operatives rather than firepower versus firepower. In short, investments in innovative and capable people, agile organisational machinery, and technology still leave funds available for acquiring more ships, submarines, tanks, aircraft, and missiles.

Acknowledging that 'arming up' is still prudent, Response Power comes with a bonus. It optimises Australia's conventional military, diplomatic, informational, and economic power when precision targeting, sharper negotiations, and immediate 'Fair Dinkum' consequences for illegal behaviour are required.

Afterthoughts

Finally, we agree with Clive Hamilton when he concludes *Silent Invasion*,

> Our naivety and our complacency are Beijing's strongest assets. Boy Scouts up against Don Corleone. But once Australians of all ethnic backgrounds understand the danger, we can begin to protect our freedoms from the new totalitarianism.

More optimistically, he concludes *Hidden Hand* with,

> People on the left and right who have opened their eyes to the threat posed by the CCP, including those who have left China to escape it, are banding together. The pushback is growing by the day, and the party bosses in Beijing are worried.

We are proud to be part of Australia's pushback. Hopefully, those in the CCP and China who seek a peaceful and prosperous world will respect Australia's efforts to de-escalate rather than defy, to disappoint rather than provoke, to continually engage positively for reconciliation rather than complain resentfully or only send a 'submarines and missiles' message over the next ten years.

Let's hope Xi Jinping and CCP hardliners conclude that a fair and respectful relationship with Australia based on trade is a better option than a grey zone escalation in 2025/26. But hope is not a strategy. Australia's de-escalation strategy and Response Power projection should be hard and hurtful enough to prompt Xi Jinping and CCP nationalists to have second thoughts about relations with Australia.

Response Power creates both disappointment and respect and leads to better relations. If not, it de-escalates and buys time before Australia's enhanced conventional military deterrence comes online in the mid-2030s and 2040s. Alan Dupont, a fellow soldier-scholar, made an insightful observation on 25 July 2020:

> Ultimately, however, we need to craft our own strategy for dealing with a more aggressive China and not rely on the goodwill of others. ... The objective of our policy is not to turn China into an enemy but to make Xi understand that while we value the relationship, we will not be coerced or threatened into compromising our sovereignty or values. ... Understanding that Xi is a devoted practitioner of hard power should inform our policy approach ...

Finally, Australia's emeritus journalist Paul Kelly warned on 4 December 2024 about what 2025 will hold:

> The world of 2025 is far different from the world of 2022. The 2020s are an age of disruption, dislocation, and destabilisation, courtesy of Donald Trump, Xi Jinping and Vladimir Putin. Nearly every certitude is under threat. Leaders must lead and smart governments get proactive. Expect fast changes to occur in today's world in a short time.

Select Bibliography

Australian Government

Australian Government 2017 *Intergovernmental agreement on Australia's national counter-terrorism arrangements,* Council of Australian Governments, Australian Government, Canberra, retrieved 2 February 2017, https://www.coag.gov.au/sites/default/files/agreements/iga-counter-terrorism.pdf

——— 2020 Royal Commission into National Natural Disaster Arrangements Report, retrieved 20 July 2023, https://naturaldisaster.royalcommission.gov.au/publications/royal-commission-national-natural-disaster-arrangements-report

Attorney-General's Department 2020 'Australia's counter-terrorism laws', Attorney General's website, retrieved 9 August 2020, https://www.ag.gov.au/national-security/australias-counter-terrorism-laws

——— 2023 *National Action Plan to Combat Modern Slavery*, Attorney-General's Department, Canberra, retrieved on 1 September 2024, https://www.ag.gov.au/crime/people-smuggling-and-human-trafficking/human-trafficking

Department of Defence 1976 *The Defence of Australia,* Defence White Paper, Department of Defence, Canberra, retrieved on 1 September 2024, https://www.aph.gov.au/About_Parliament/Parliamentary_Departments/Parliamentary_Library/pubs/rp/rp1516/DefendAust/1976

——— 1987 *Defence White Paper*, Department of Defence, Canberra, retrieved 1 September 2024 https://www.defence.gov.au/sites/default/files/2021-08/wpaper1987.pdf

——— 1994 *Defence White Paper*, Department of Defence, Canberra, retrieved on 1 September 2024, https://www.aph.gov.au/About_Parliament/Parliamentary_departments/Parliamentary_Library/pubs/rp/rp1516/DefendAust/1994#:~:text=The%201994%20Defence%20White%20Paper%20commended%20the%20success%20of%20the,oriented%20Australian%20manufacturing%20and%20services

——— 1995 *Serving Australia, The Australian Defence Force in the Twenty-First Century (The Glenn Review), Australian* Government, Canberra, retrieved on 1 September 2024, https://www.defence.gov.au/about/strategic-planning/defence-white-paper

——— 2000 *Defence White Paper*, Department of Defence, https://www.aph.gov.au/About_Parliament/Parliamentary_Departments/Parliamentary_Library/pubs/rp/rp1516/DefendAust/2000

——— 2001 *Defence Annual Report 2000–01.* Australian Government, Canberra, retrieved 1 September 2024, https://www.defence.gov.au/about/accessing-information/annual-reports

——— 2003 *Government response to the Joint Standing Committee on Foreign Affairs, Defence and Trade Report 'From Phantom to Force, Towards a More Efficient and Effective Army' and subsequent report 'A Model for a New Army, Community Comments on the 'From Phantom to Force' Parliamentary Report into the Army'*, Department of Defence, Canberra, retrieved 1 September 2024, https://parlinfo.aph.gov.au/parlInfo/search/display/display.w3p;query=Id:%22library/lcatalog/00118977%22

——— 2003 Department of Defence Annual Report 2002–03, retrieved 28 April 2023, https://www.defence.gov.au/about/accessing-information/annual-reports

——— 2003 *Future Warfighting Concept*, Canberra, retrieved 1 September 2019, https://www.defence.gov.au/publications/fwc.pdf

——— 2005 Strategic Update 2005, Australian Government, Canberra, retrieved 1 September 2024, https://www.aph.gov.au/About_Parliament/Parliamentary_departments/Parliamentary_Library/pubs/rp/rp1516/DefendAust/NationalSecurity

——— 2007 Report of the Defence Management Review, Australian Government, Canberra, retrieved 28 April 2023, https://www.defence.gov.au/about/accessing-information/annual-reports

——— 2007 *Australia's National Security – A Defence Update 2007*, Canberra, retrieved 1 September 2024, https://www.aph.gov.au/About_Parliament/Parliamentary_departments/Parliamentary_Library/pubs/rp/rp1516/DefendAust/NationalSecurity

——— 2007 *Annual Report 2006–07*, Volume One, Australia Government, Canberra, retrieved 28 April 2023 https://www.defence.gov.au/about/accessing-information/annual-reports

——— 2008 *Annual Report 2007–08*, Volume One, Australia Government, Canberra, retrieved 28 April 2023 https://www.defence.gov.au/about/accessing-information/annual-reports

——— 2009 *2008 Audit of the Defence Budget* (Pappas Report). Australian Government, Canberra

——— 2009 Defence Response to the Defence Budget Audit, 17 November 2009, Australian Government, Canberra

——— 2009, *Annual Report 2008–09*, Volume One, Australia Government, Canberra, retrieved 28 April 2023, https://www.defence.gov.au/about/accessing-information/annual-reports

——— 2009 *The Strategic Reform Plan – Delivering Force 2030*, Canberra, retrieved 1 September 2024, https://www.aph.gov.au/About_Parliament/Parliamentary_departments/Parliamentary_Library/pubs/BriefingBook43p/defencesrp#:~:text=In%20an%20April%202010%20publication,%2C%20to%20'fully%20mature'

——— 2010 *Counter-Terrorism White Paper- Securing Australia, Protecting Our Community*, Australian Government, Canberra, retrieved 1 February 2019, https://

www.dst.defence.gov.au/sites/default/files/basic_pages/documents/counter-terrorism-white-paper.pdf

——— 2010 *The Strategic Reform Program, Making It Happen*, retrieved 1 September 2024, https://www.aph.gov.au/About_Parliament/Parliamentary_departments/Parliamentary_Library/pubs/BriefingBook43p/defencesrp#:~:text=In%20an%20April%202010%20publication,%2C%20to%20'fully%20mature'

——— 2011. *Report of the Australian Defence Force Personal Conduct Review – Beyond Compliance, Professionalism, Trust and Capability in the Australian Profession of Arms,* Australian Government, Canberra, retrieved 1 September 2024, https://www.legal-tools.org/doc/a4486a/pdf/

——— 2011 *Annual Report 2010–11*, Volume One, Australia Government, Canberra, retrieved 28 April 2023, https://www.defence.gov.au/about/accessing-information/annual-reports

——— 2012 *Pathways to Change, Evolving Defence Culture*, Australian Government, Canberra, retrieved on 1 September 2024, https://www.defence.gov.au/about/reviews-inquiries/pathway-change-evolving-defence-culture

——— 2012 *Defence Assistance to the Civil Community Manual,* Canberra, Australian Defence Force, Canberra, retrieved on 1 September 2024, https://www.defence.gov.au/defence-activities/programs-initiatives/defence-assistance-civil-community-initiative

——— 2013 Assistant Minister for Defence – Launch of Project Suakin by Stuart Robert at HMAS Harman, 26 November 2013, Canberra, retrieved on 1 September 2024, https://www.minister.defence.gov.au/transcripts/2013-11-26/assistant-minister-defence-launch-project-suakin-stuart-robert-hmas-harman

——— 2012 *Australian Defence Force Posture Review*, Canberra, retrieved on 1 September 2024, https://www.defence.gov.au/about/reviews-inquiries/adf-posture-review

——— 2014 Annual Report 2013–14 – Volume One – Performance, governance and accountability, retrieved 04 September 2023, https://www.defence.gov.au/about/accessing-information/annual-reports

——— 2014 *Annual Report 2013–14*, Volume Two, Australia Government, Canberra, https://www.defence.gov.au/about/accessing-information/annual-reports

——— 2014 *Defence Diversity and Inclusion Strategy 2012–2017*, Australian Government, Canberra, retrieved on 1 September 2024, https://www.defence.gov.au/jobs-careers/defence-aps-jobs/what-defence-offers/diversity-inclusion

——— 2015 *Creating One Defence, First Principles Review*, Australian Government, Canberra, retrieved on 1 September 2024, https://www.defence.gov.au/about/reviews-inquiries/first-principles-review-creating-one- defence#:~:text=The%20First%20Principles%20Review%20was,and%20become%20an%20integrated%20organisation

——— 2015 Department of Defence Annual Report 2014–15, retrieved on 1 September 2024, https://www.defence.gov.au/about/accessing-information/annual-reports

——— 2016 *Future Operating Environment 2035*, Australian Government: Canberra, retrieved 1 September 2024, https://cove.army.gov.au/sites/default/files/08-09_0/08/Future-Operating-Environment-2035.pdf

——— 2016 *Defence White Paper*, Department of Defence, Canberra, retrieved 22 June 2024, https://www.defence.gov.au/about/strategic-planning/defence-white-paper

——— 2017 Army delivers final component of Plan Beersheba, 28 October 2017, retrieved 06 April 2023, https://www.defence.gov.au/news-events/releases/2017-10-28/army-delivers-final-component-plan-beersheba

——— 2017 *The Strategy Framework*, Department of Defence, Canberra, retrieved on 1 September 2024, https://www.defence.gov.au/about/strategic-planning

——— 2018 Answer to question on notice, QON127, SSCFADT – AE – 28 Feb 18 – Q127 – Reserves – Gallacher, Senate Standing Committee on Foreign Affairs and Trade, retrieved 5 April 2023, https://www.aph.gov.au/Parliamentary_Business/Senate_Estimates/eqon

——— 2018 *Reserve Remuneration Industrial History (As at 2017), A potted history*. Defence People Group, Canberra, retrieved on 1 September 2024, https://dra.org.au/conference-2018-item/32881/review-of-adf-sercat-3-5-reserve-employment-package/?type_fr=901

——— 2019 *Department of Defence Annual Report 2018–19*, retrieved 04 September 2023, retrieved on 1 September 2024, https://www.defence.gov.au/about/accessing-information/annual-reports

——— 2019 *Defence Census 2019 Public Report*, Australian Government, Canberra, retrieved on 1 September 2024, https://www.defence.gov.au/about/accessing-information/defence-census

——— 2019 *Guidelines for requesting support from the Australian Defence Force (Call-out) under Part IIIAAA of the Defence Act 1903 by States and Territories*, 10 June 2019, Department of Defence, Canberra, retrieved on 1 September 2024, https://www.aph.gov.au/Parliamentary_Business/Bills_Legislation/bd/bd1819a/19bd043

——— 2020, *Future Submarine Program – Transition to Design*, Department of Defence, retrieved 23 January 2020, https://www.anao.gov.au/work/performance-audit/future-submarine-program-transition-to-design

——— 2020, *Joint Strike Fighter – Introduction into Service and Sustainment Planning, Report No 14*, Department of Defence, Summary, retrieved 24 January 2020, https://www.anao.gov.au/work/performance-audit/joint-strike-fighter-introduction-service-and-sustainment

——— 2020 *Defence Strategic Update*, Department of Defence, retrieved 23 June 2024, https://www.defence.gov.au/about/strategic-planning/2020-defence-strategic-update

——— 2020 *Defence Force Structure Update*, Department of Defence, retrieved 23 June 2024, https://www.defence.gov.au/about/strategic-planning/2020-force-structure-plan#:~:text=The%202020%20Force%20Structure%20Plan,the%20 2020%20Defence%20Strategic%20Update

——— 2020 Morrison Government invests in additional Australian-made soldier capability for Defence, Media Release, 13 July 2020, retrieved 03 August 2023, https://www.minister.defence.gov.au/media-releases/2020-07-13/morrison-government-invests-additional-australian-made-soldier-capability-defence

——— 2020 'Operation Bushfire Assist 2019–2020 Daily Update', 23 January, retrieved 23 January February 2020, https://www.minister.defence.gov.au/media-releases/2020-03-26/operation-bushfire-assist-concludes

——— 2022 Department of Defence Annual Report 2021–22, retrieved 18 July 2023, https://www.defence.gov.au/about/accessing-information/annual-reports

——— 2022 *2021–2040 Defence Strategic Workforce Plan – Part 1*, Australian Government, Canberra, retrieved on 1 September 2024, https://www.transparency. gov.au/publications/defence/department-of-defence/department-of-defence-annual-report-2021-22/chapter-6---strategic-workforce-management/workforce-planning

——— 2022 *2021–2040 Defence Strategic Workforce Plan -Part 2*, Australian Government, Canberra, retrieved on 1 September 2024, https://www.transparency. gov.au/publications/defence/department-of-defence/department-of-defence-annual-report-2021-22/chapter-6---strategic-workforce-management/workforce-planning

——— 2023 ADF Employment Offer Modernisation Program, retrieved 05 May 2023, https://pay-conditions.defence.gov.au/adf-employment-modernisation

——— 2023, *National Defence: Defence Strategic Review*, Department of Defence, retrieved 23 June 2024, https://www.defence.gov.au/about/reviews-inquiries/ defence-strategic-review

——— 2024, National Defence Strategy and Integrated Investment Plan, Department of Defence, retrieved 23 June 2024, https://www.defence.gov.au/about/strategic-planning/2024-national-defence-strategy-2024-integrated-investment-program

——— 'Global Operations', Defence Department website, retrieved 7 June 2020, https://www.defence.gov.au/operations/

Department of External Affairs, *Security Treaty between Australia, New Zealand and the United States of America [ANZUS]*, Australian Treaty Series 1952 No 2, San Francisco, 1 September 1952, Article IV, retrieved 14 February 2021, http://www. austlii.edu.au/au/other/dfat/treaties/1952/2.html

Department of Foreign Affairs and Trade 2017, *2017 Foreign Policy White Paper,* *Canberra*, retrieved 19 February 2021, https://www.dfat.gov.au/publications/ minisite/2017-foreign-policy-white-paper/fpwhitepaper/pdf/2017-foreign-policy-white-paper.pdf

——— 2020, *Australia's Pacific engagement, Stepping Up Australia's engagement with our Pacific family*, DFAT website, retrieved 20 June 2020

Department of Home Affairs 2010 *Counter-Terrorism White Paper- Securing Australia, Protecting Our Community*, Australian Government, Canberra, retrieved 1 February 2019, https://www.dst.defence.gov.au/sites/default/files/basic_pages/documents/counter-terrorism-white-paper.pdf

——— 2015, *Review of Australia's Counter-Terrorism Machinery*, Home Affairs Department, Canberra, retrieved 20 March 2020, https://www.homeaffairs.gov.au/nat-security/files/review-australia-ct-machinery.pdf

——— 2018 *Annual Report, 2017–18*, Department of Home Affairs, Canberra, retrieved 11 August 2019, https://www.afp.gov.au/annual-report-2017-18

——— 2019 *Australian Public Service Responsibilities for Managing Climate and Disaster Risk*, Canberra, Disaster and Climate Resilience Reference Group

——— 2020 'Our History', Home Affairs website, retrieved 20 July 2020, https://www.homeaffairs.gov.au/about-us/who-we-are/our-history

——— 2020 website Counter Foreign Interference, retrieved 12 August 2020, https://www.homeaffairs.gov.au/about-us/our-portfolios/national-security/countering-foreign-interference/cfi-strategy

Department of Prime Minister and Cabinet 2013 *Strong and Secure: A Strategy for Australia's National Security*, Commonwealth of Australia, Canberra, retrieved 23 June 2024, https://apo.org.au/node/33996

——— 2015, *National Cyber Security Strategy*, Department of Prime Minister and Cabinet Canberra, retrieved 1 September 2024, https://www.homeaffairs.gov.au/cyber-security-subsite/files/PMC-Cyber-Strategy.pdf

——— 2015 *Financing Terrorism*, Department of Prime Minister and Cabinet, Canberra, retrieved on 1 September 2024, https://www.aph.gov.au/About_Parliament/Parliamentary_Departments/Parliamentary_Library/pubs/BriefingBook45p/MoneyLaundering

——— 2015 *Martin Place siege: Joint Commonwealth-New South Wales Review*, Department of Prime Minister and Cabinet, Canberra, retrieved 1 September 2024, https://www.homeaffairs.gov.au/nat-security/files/martin-place-siege-nsw-review.pdf

——— 2015 *National Counter Terrorism Strategy- Crowded Places*, Department of Prime Minister and Cabinet, Canberra, retrieved 1 September 2024, https://www.nationalsecurity.gov.au/crowded-places-subsite/Files/australias-strategy-protecting-crowded-places-terrorism.pdf

——— 2015 *National Counter Terrorism Strategy- Active Shooter*, Department of Prime Minister and Cabinet, Canberra, retrieved 1 September 2024, https://www.homeaffairs.gov.au/about-us/our-portfolios/national-security/countering-extremism-and-terrorism/centre-for-counter-terrorism-coordination

——— 2015 *National Counter Terrorism Strategy- Improvised Explosive Device*, Department of Prime Minister and Cabinet, Canberra, retrieved 1 September 2024,

https://www.homeaffairs.gov.au/about-us/our-portfolios/national-security/
countering-extremism-and-terrorism/centre-for-counter-terrorism-coordination

——— 2015 *National Counter Terrorism Strategy- Chemical Weapons,* Department
of Prime Minister and Cabinet, Canberra., retrieved 1 September 2024, https://
www.homeaffairs.gov.au/about-us/our-portfolios/national-security/countering-
extremism-and-terrorism/centre-for-counter-terrorism-coordination

——— 2015 *National Cyber Security Strategy,* Canberra, retrieved 1 September 2024,
https://www.homeaffairs.gov.au/about-us/our-portfolios/national-security/
countering-extremism-and-terrorism/centre-for-counter-terrorism-coordination

——— 2015 *Protection of National Infrastructure,* Department of Prime Minister
and Cabinet, Canberra, retrieved on 1 September 2024, https://www.homeaffairs.
gov.au/about-us/our-portfolios/national-security/security-coordination/critical-
infrastructure-resilience

——— 2017, *2017 Independent Intelligence Review*, Canberra, retrieved 12 May 2020,
https://www.pmc.gov.au/national-security/2017-independent-intelligence-review

——— 2023, *Australian Government Crisis Management Framework*, Version 3.3
September 2023, Canberra, retrieved 2 September 2024, https://www.pmc.
gov.au/sites/default/files/resource/download/australian-government-crisis-
management-framework_0.pdf

——— 2024 *National Counter Terrorism Plan*, 5th Edition, Department of Prime
Minister and Cabinet, Canberra, retrieved on 1 September 2024, https://www.
nationalsecurity.gov.au/what-australia-is-doing-subsite/Files/anzctc-national-
counter-terrorism-plan.PDF

Department of Administrative Services, Mark, R 1978, *Report to the Minister
for Administrative Services on the Organisation of Police Resources
in the Commonwealth Area and Other Related Matters*, Australian
Government Publishing Service, Canberra, retrieved on 1 September
2024, https://parlinfo.aph.gov.au/parlInfo/search/display/display.
w3p;query=Id%3A%22publications%2Ftabledpapers%2FHPP052016005700%22

Australian Government Agencies

Australian Bureau of Statistics 2021 *Cultural diversity of Australia*, retrieved 3 October
2023, https://www.legislation.gov.au/Details/C2021C00371 https://www.abs.gov.
au/articles/cultural-diversity-australia

——— 2023 *Permanent migrants in Australia*, 29 March 2023, retrieved 20 August
2023 https://www.abs.gov.au/statistics/people/people-and-communities/
permanent-migrants-australia/latest-release

Australian Defence Force 2013 Preparedness and Mobilisation Executive, Australian
Defence Doctrine Publication (ADDP), 00.2, Defence Intranet, Defence Publishing
Service, Canberra

——— 2020, *Information Warfare–ADF Manoeuvre in the Information Environment,*
Commonwealth of Australia Publishing, Canberra

——— 2020 'Review of OP Bushfire Assist 19–20 – Initial Department Insights', *Noting Brief for the Secretary and CDF*, Defence intranet, Department of Defence, Canberra

——— 2020 *Chief of the Defence Force Order of the Day – Revocation of the Reserve Call-out Order*, 12 February 2020, Defence intranet, Canberra, Department of Defence

——— 2022 *Aspire: The Australian Defence Force Theatre Concept*, Defence intranet, Department of Defence: Canberra

——— 2024 *Australian Defence Glossary*, Defence intranet, Department of Defence, Canberra

Australian National Archives, Official History of the Operations and Administration of Special Operations – Australia [(SOA), circa 1948, also known as the Inter-Allied Services Department (ISD) and Services Reconnaissance Department (SRD)] Volume 2 – Operations – copy no 4 [for Director, Military Intelligence (DMI), Headquarters (HQ), Australian Military Forces (AMF), Melbourne – abridged version of copy no 1]. Series no. A3269, O8/B, National Australian Archives Canberra, retrieved 31 August 2020, https://recordsearch.naa.gov.au/SearchNRetrieve/Interface/ViewImage.aspx?B=235326

Australian National Audit Office, 1997 *Army Presence in the North, Department of Defence Performance Audit*, Audit Report No 27 1996–97, ANAO, Canberra, March 1997, retrieved 1 September 2023 https://www.anao.gov.au/work/performance-audit/army-presence-the-north

——— 2000, Commonwealth Emergency Management Arrangements, Audit Report No 41, 28 April, retrieved 21 June 2024 https://www.anao.gov.au/work/performance-audit/commonwealth-emergency-management-arrangements

——— 2001 *Australian Defence Force Reserves*, Audit Report No.33 2000–2001, ANAO, Canberra, retrieved 22 June 2024, https://www.anao.gov.au/work/performance-audit/australian-defence-force-reserves

——— 2002 *Management of Australian Defence Force Deployments to East Timor*, Audit Report No. 38 2001–02, The Auditor General, Canberra, retrieved 22 June 2024, https://www.anao.gov.au/pubs/annual-reports?query=%3East+Timor&items_per_page=10

——— 2005 *Army Capability Assurance Processes*, Audit Report No. 25 2004–05, The Auditor General, Canberra, retrieved 22 June 2024, https://www.anao.gov.au/work/performance-audit/army-capability-assurance-processes

——— 2009 *Australian Defence Force Reserves*, Audit Report No.31 2008–09, The Auditor General, Canberra, retrieved 1 September 2024, https://www.anao.gov.au/sites/default/files/ANAO_Report_2008-2009_31.pdf

——— 2014 *Emergency Defence Assistance to the Civil Community*, Audit Report No. 24, 2013–14 The Auditor General, Canberra. (Canberra, ANAO), retrieved 1 September 2024, https://www.anao.gov.au/sites/default/files/AuditReport_2013-2014_24.pdf

Australian Federal Police 2014, *Annual Report, 2014–15*, Department of Home Affairs, Canberra, retrieved 11 August 2019, https://www.afp.gov.au/afp-annual-report-2014-15#chapter_4-2

——— 2021 Australian Federal Police Organisation Structure, AFP website, afp.gov. au, retrieved 15 February 2021, https://www.afp.gov.au/sites/default/files/PDF/AFPOrgStructure-08022021.pdf

——— 2021, Specialist Response Group, AFP website, afp.gov.au, retrieved 15 February 2021, https://www.afp.gov.au/what-we-do/operational-support/specialist-response-group

Australian New Zealand Counter-Terrorism Committee 2017, *National Counter Terrorism Plan*, 4th ed, retrieved 23 March 2021, https://www.nationalsecurity.gov.au/Media-and-publications/Publications/Documents/ANZCTC-National-Counter-Terrorism-Plan.PDF

Australian Human Rights Commission, 2020, *International Covenant on Civil and Political Rights – Human Rights at your Fingertips*, AHRC website, retrieved 16 July 2020 https://humanrights.gov.au/our-work/commission-general/international-covenant-civil-and-political-rights-human-rights-your

——— 2020, *Convention against Torture and Other Cruel, Inhuman and Degrading Treatment or Punishment – Human Rights at your Fingertips*, AHRC website, retrieved 16 July 2020, https://humanrights.gov.au/our-work/commission-general/convention-against-torture-and-other-cruel-inhuman-or-degrading

Australian Secret Intelligence Service, Burgess, M 2020, 'Director-General's Annual Threat Assessment 2020', Australian Security Intelligence Organisation website, 24 February, retrieved 23 May 2020, https://www.asio.gov.au/publications/speeches-and-statements/director-general-annual-threat-assessment-0.html

——— 2020, Counter Espionage and Foreign Interference, ASIO website, retrieved 12 August 2020, https://www.asio.gov.au/counter-espionage.html

——— Burgess, M 2021, 'Director-General's Annual Threat Assessment 2021', Australian Security Intelligence Organisation website, 17 March, retrieved 21 June 2024, https://www.asio.gov.au/resources/speeches-and-statements/director-generals-annual-threat-assessment-2021

——— Burgess, M 2022, 'Director-General's Annual Threat Assessment 2022', Australian Security Intelligence Organisation website, 9 February, retrieved 21 June 2024, https://www.asio.gov.au/resources/speeches-and-statements/director-generals-annual-threat-assessment-2022

——— Burgess, M 2023, 'Director-General's Annual Threat Assessment 2023', Australian Security Intelligence Organisation website, 21 February, retrieved 21 June 2024, https://www.asio.gov.au/director-generals-annual-threat-assessment-2023

——— Burgess, M 2024 'Director-General's Annual Threat Assessment 2024', Australian Security Intelligence Organisation website, 28 February, retrieved

21 June 2024 https://www.asio.gov.au/director-generals-annual-threat-
assessment-2024

Australian Government Legislation

Parliament of Australia 1901, *The Constitution*, Australian Government, Melbourne,
retrieved 26 May 2018, retrieved 26 May 2018, https://www.legislation.gov.au/
Details/C2005Q00193

——— 1903, *Defence Act* (Cth), Commonwealth of Australia, Canberra, retrieved
5 June 2020, https://www.legislation.gov.au/C1903A00020/latest/text

——— 1995 *Criminal Code Act 1995*, Australian Government, Canberra, retrieved
26 May 2018, retrieved 26 May 2018, https://www.legislation.gov.au/Search/
C2018C00386%20retrieved%20May%2026%202018

——— 2001 *Border Protection Act,* (Cth), Commonwealth of Australia, Canberra,
retrieved 1 September 2024, https://www.aph.gov.au/Parliamentary_Business/
Bills_Legislation/Bills_Search_Results/Result?bId=r1412

——— 2001 *Intelligence Services Act,* (Cth), Commonwealth of Australia, Canberra,
retrieved 1 September 2024, https://www.aph.gov.au/Parliamentary_Business/
Bills_Legislation/Bills_Search_Results/Result?bId=r1350

——— 2002, *Report – Select Committee for an Inquiry into a certain maritime
incident,* CanPrint Communications for Senate Printing Unit, Commonwealth of
Australia, Canberra, retrieved 21 May 2020, https://www.aph.gov.au/binaries/
senate/committee/maritime_incident_ctte/report/report.pdf

——— 2011, *Human Rights (Parliamentary Scrutiny) Act 2011*, retrieved 16 July
2020, https://www.legislation.gov.au/Details/C2011A00186

——— 2012 *ASIO Act,* (Cth), Commonwealth of Australia, Canberra, retrieved on
1 September 2024, https://www.legislation.gov.au/C2004A02123/2013-06-29/
text

——— 2018, *National Security Legislation Amendment (Espionage And Foreign
Interference) Bill 2017, Explanatory Memorandum*, Circulated by authority of
the Attorney-General, Senator the Honourable George Brandis QC, 29 June,
retrieved 20 August 2020, https://www.aph.gov.au/Parliamentary_Business/
Bills_Legislation/Bills_Search_Results/Result?bId=r6022

——— 2018, *National Security Legislation Amendment (Espionage and Foreign
Interference) Act 2018*, No. 67, Canberra, retrieved 14 July 2020, https://www.
legislation.gov.au/Details/C2018A00067

——— 2018, *Introduction of the Defence Amendment (Call Out of the Australian
Defence Force) Bill 2018*, 28 June, Parliament House, Canberra, retrieved
1 September 2024, https://www.aph.gov.au/Parliamentary_Business/Bills_
LEGislation/Bills_Search_Results/Result?bId=r6149#:~:text=Implements%20
certain%20recommendations%20of%20the,enable%20call%20out%20
orders%20to

—— 2018, *Counter-Terrorism Legislation Amendment Bill (No. 1) 2018, Explanatory Memorandum, Schedule 1 Amendments Current Threat Environment,* Federal *Register of Legislation,* retrieved 23 June 2024, https://www.aph.gov.au/ Parliamentary_Business/Bills_Legislation/bd/bd1819a/19bd004

—— 2018, *Call Out of the Australian Defence Force Bill 2018,* Australian Government, Canberra, retrieved 20 May 2020, https://www.legislation.gov.au/ Details/C2018A00158

—— 2018, 'National Security Legislation Amendment (Espionage and Foreign Interference) Act 2018', C2018A00067, 29 June, Federal Register of Legislation, retrieved 9 August 2020, https://www.legislation.gov.au/Details/C2018A00067

—— 2020 *Australia's Foreign Relations (State and Territory Arrangements) Bill 2020 and Australia's Foreign Relations (State and Territory Arrangements) (Consequential Amendments) Bill 2020,* retrieved 20 October 2020, https://www. aph.gov.au/Parliamentary_Business/Committees/Senate/Foreign_Affairs_Defence_ and_Trade/AustForeignRelations2020

Attorney General of the Parliament of Australia 2018 *Passage of the Defence Amendment (Call out of the Australian Defence Force) Bill 2018,* media release, 27 November, retrieved July 15 2019, https://www.attorneygeneral.gov.au/Media/ Pages/Passage-of-the-Defence-Amendment-Call-out-of-The-Australian-Defence-Force-Bill-2018-27-november-2018.aspx

—— 2015 *Australian Border Force Act 2015 (Cth),* Commonwealth of Australia, Canberra, retrieved on 1 September 2024, https://www8.austlii.edu.au/cgi-bin/ viewdb/au/legis/cth/consol_act/abfa2015225/

—— 2017 *2017 Independent Intelligence Review,* Department of Prime Minister and Cabinet, Canberra, retrieved 12 May 2020, https://www.pmc.gov.au/national-security/2017-independent-intelligence-review

—— 2018 *Introduction of the Defence Amendment (Call Out of the Australian Defence Force) Bill 2018,* 28 June, Parliament House, Canberra

—— 2018 *National Security Legislation Amendment (Espionage and Foreign Interference) Act, amendment to Criminal Code Act 1995 (Cth),* Commonwealth of Australia, Canberra

—— 2018 *The Defence Amendment (Call out of the Australian Defence Force) Bill 2018,* Parliament of Australia, Canberra, retrieved 9 April 2020, https://www. legislation.gov.au/Details/C2018A00158

—— 2019, *Foreign Influence Transparency Scheme Amendment Bill amendment to Foreign Influence Transparency Scheme Act 2018,* Commonwealth of Australia, Canberra

Government of New South Wales

New South Wales Government 2002, *Terrorism (Police Powers) Act 2002 No 115,* New South Wales Government, Sydney, retrieved 16 June 2018, https://www.legislation.nsw.gov.au/inforce/6f5a8912-c4e1-e5ba-aa92-93c10aa399a9/2002-115.pdf

Government of Victoria

State of Victoria 2010, R v Vinayagamoorthy & Ors. VSC 148, 31 March, Melbourne

State of Victoria 2017, Victorian Coroner's Inquest into the death of Yacqub Khayre, Melbourne

State of Victoria 2017, Justice Assurance and Review Office's review of Corrections Victoria's management of Khayre (JARO Review), Justice Assurance and Review Office, Melbourne

State of Victoria 2017, Expert Panel on Terrorism Report 2: The second report on how Victoria's legislation, powers and procedures are working to prevent, monitor, investigate and respond to terrorism, Department of Premier and Cabinet, Melbourne

Government of Western Australia

Western Australian Government, 2008, Billing v The State of Western Australia, WASCA, 21 January, Perth

Government of Canada

Eggleton A (2000) *Government of Canada Policy Statement Land Force Reserve Restructure, Minister of National Defence*, 12 October, Government of Canada, Alberta

Government of Estonia

Estonian Government, 2017, 'How Estonia Became a Global Heavyweight in Cyber Security', E-Estonia, Estonian Government, retrieved 21 October 2020 at https://e-estonia.com/

Estonian Government 2020, 'Estonian Demographics: 2020 Population', *Statistics Estonia*, Estonian Government, retrieved 20 October 2020 at http://www.stat.ee

NZ Government

New Zealand Ministry of Foreign Affairs and Trade, 'Building on a Success Story', MFAT website, retrieved 17 February 2021, https://www.mfat.govt.nz/en/trade/free-trade-agreements/free-trade-agreements-concluded-but-not-in-force/nz-china-free-trade-agreement-upgrade/overview/#:~:text=The%20New%20Zealand%2DChina%20Free,success%20story%20for%20both%20countries.&text=China%20is%20now%20New%20Zealand's,free%20trade%20agreement%20was%20signed

Australian New Zealand Counter-Terrorism Committee 2017, *National Counter Terrorism Plan*, 4th ed, p. 19, pp. 23–24, retrieved 23 March 2021, https://www.nationalsecurity.gov.au/Media-and-publications/Publications/Documents/ANZCTC-National-Counter-Terrorism-Plan.PDF

Government of Japan

Ministry of Foreign Affairs of Japan, 1960, Treaty of Mutual Cooperation and Security between Japan and the United States of America, 19 January, retrieved on 14 February 2021, https://www.mofa.go.jp/region/n-america/us/q&a/ref/1.html

US Government

US Government, Kean, T.H, Hamilton, LH, Ben-Veniste, R, Kerrey, B, Fielding, FF, Lehman, JF, Gorelick, JS, Roemer, TJ, Gorton, S, & Thompson, JR 2004. *The 9–11 Commission' Report, National Commission on Terrorist Attacks Upon the United States*, United States Government, Washington DC, p. 346, retrieved 18 May 2020, https://www.9-11commission.gov/report/911Report.pdf

Department of Defense 2001 *2001 Quadrennial Defense Review Report*, Department of Defense, Washington, retrieved 2 September 2024, https://history.defense.gov/Historical-Sources/Quadrennial-Defense-Review/

——— 2006 *2006 Quadrennial Defense Review Report*, Department of Defense, Washington, retrieved 2 September 2024, https://history.defense.gov/Historical-Sources/Quadrennial-Defense-Review/

——— 2008 *Transforming the National Guard and Reserves into a 21st-Century Operational Force*, Commission on the National Guard and Reserves, Final Report to Congress and the Secretary of Defense, retrieved on 2 September 2024, https://policy.defense.gov/portals/11/Documents/hdasa/references/CNGR%20Final%20Report.pdf

——— 2010 *2010 Quadrennial Defense Review Report*, Department of Defense, Washington, retrieved 2 September 2024, https://history.defense.gov/Historical-Sources/Quadrennial-Defense-Review/

——— 2011 *Comprehensive Review of the Future Role of the Reserve Component*, Volume I Executive Summary & Main Report, Office of the Vice Chairman of the Joint Chiefs of Staff, Washington, retrieved on 2 September 2024

——— 2014 *2014 Quadrennial Defense Review Report*, Department of Defense, Washington, retrieved 2 September 2024, https://history.defense.gov/Historical-Sources/Quadrennial-Defense-Review/

——— 2014, *Strategy for countering weapons of mass destruction*, United States Special Operations Command, Tampa, https://dod.defense.gov/Portals/1/Documents/pubs/DoD_Strategy_for_Countering_Weapons_of_Mass_Destruction_dated_June_2014.pdf

——— 2017, *Russia Military Power: Building a Military to Support Great Power Aspirations*, Defense Intelligence Agency, Government Printing Office, Washington DC, retrieved on 4 May 2020, https://www.dia.mil/Military-Power-Publications/

——— 2018, *National Strategy for Counter-terrorism- October 2018*, The White House National Security Committee, Washington DC, retrieved 11 March 2019, https://www.whitehouse.gov/wp-content/uploads/2018/10/NSCT.pdf_

——— 2019, *China Military Power: Modernizing a Force to Fight and Win*, Defense Intelligence Agency, Government Printing Office, Washington DC, retrieved on 4 May 2020, https://www.dia.mil/Military-Power-Publications/

Central Intelligence Agency, Federal Bureau of Investigation, and National Security Agency 2017, Assessing Russian Activities and Intentions in Recent US Elections, *Intelligence Community Assessment*, 6 January, retrieved 4 February, https://www.dni.gov/files/documents/ICA_2017_01.pdf

US Embassies and Consulates in China, 2019, '2018 Report on International Religious Freedom: China (Includes Tibet, Xinjiang, Hong Kong, and Macau)', 21 June 2019, retrieved 14 July 2020, https://china.usembassy-china.org.cn/2018-report-on-international-religious-freedom-china/

US Army 2017, *Multi-Domain Battle: Evolution of Combined Arms for the 21st Century 2025–2040,* Training United States Army 2011, Army Doctoral Publication 3-05 Special Operations, Government Printing Office, Washington, DC, Doctrine Command, Fort Eustis, VA, retrieved 12 September 2020 https://www.tradoc.army.mil/Portals/14/Documents/MDB_Evolutionfor21st%20(1).pdf

——— 2018 *The Army Strategy*, retrieved 20 April 2020, https://www.ausa.org/news/army-leaders-release-new-army-strategy

——— 2019 *The US Army in Multi-Domain Operations in 2028*. Department of Defense: Washington, retrieved on 2 September 2024, https://www.army.mil/article/243754/the_u_s_army_in_multi_domain_operations_2028

US Special Operations Command, JP 3-05, 2013, Special Operations, United States Special Operations Command, Tampa, retrieved 23 June 2024, https://apps.dtic.mil/sti/citations/ADA543873

——— 2015 *The Grey Zone*, White Paper, dated 9 September 2023, retrieved 1 August 2023, https://info.publicintelligence.net/USSOCOM-GrayZones.pdf

US Citizenship and Immigration Services 2023 *Naturalisation Through Military Service*, retrieved 20 August 2023, https://www.uscis.gov/military/naturalization-through-military-service

UK Government

UK Government 1994 Front Line First, The Defence Costs Study 14/7/94. UK Government, London, retrieved on 2 September 2024, https://researchbriefings.files.parliament.uk/documents/RP94-101/RP94-101.pdf

——— 2003 *Defence White Paper – Delivering Security in a Changing World,* Ministry of Defence, London, retrieved on 2 September 2024, https://publications.parliament.uk/pa/cm200304/cmselect/cmdfence/465/465.pdf

——— 2003 *Defence White Paper – Supporting Essays,* Ministry of Defence, London, retrieved on 2 September 2024, https://publications.parliament.uk/pa/cm200304/cmselect/cmdfence/465/465.pdf

——— 2008 'Eighth Report, The historical context of the fall of the Soviet Union, Options for Change and Front Line First', Select Committee on Defence, retrieved

01 August 2023, https://publications.parliament.uk/pa/cm199798/cmselect/cmdfence/138/13805.htm

———— 2010 *Securing Britain in an Age of Uncertainty, The Strategic Defence and Security Review*, Ministry of Defence, London, retrieved on 2 September 2024, https://www.gov.uk/government/publications/the-strategic-defence-and-security-review-securing-britain-in-an-age-of-uncertainty

———— 2014, 'The situation in Iraq and Syria and the response to al-Dawla al-Islamiya fi al-Iraq al-Sham (DAESH)', House of Commons Defence Committee, 7th Report of Session, 15, London, retrieved on 2 September 2024, https://committees.parliament.uk/work/2212/the-situation-in-iraq-and-syria/publications/

———— 2017 Understanding Hybrid Warfare, Countering Hybrid Warfare (CHW), Multinational Capability Development Campaign (MCDC) Project, UK Government, London, retrieved 12 February 2020, https://www.gov.uk/government/publications/countering-hybrid-warfare-project-understanding-hybrid-warfare

———— 2018, *CONTEST- the United Kingdom's Strategy for Countering Terrorism-June 2018,* Home Department, London, retrieved 1 February 2019, https://www.safecampuscommunities.ac.uk/uploads/files/2019/03/contest_uks_counter_terrorism_strategy.pdf

———— 2019, Countering Hybrid Warfare, Countering Hybrid Warfare, (CHW) Multinational Capability Development Campaign (MCDC) Project, UK Government, London, retrieved 12 February 2020, https://www.gov.uk/government/publications/countering-hybrid-warfare-project-understanding-hybrid-warfare

House of Commons Defence Committee 2014 *Review of) Future Army 2020*, UK Government, London, retrieved on 2 September 2024, https://publications.parliament.uk/pa/cm201314/cmselect/cmdfence/576/57607.htm

House of Lords Library 2008 *Library Note, Reserve Forces* (LLN 2008/014) UK Government, London, retrieved on 2 September 2024, https://researchbriefings.files.parliament.uk/documents/LLN-2008-014/LLN-2008-014.pdf

UK Ministry of Defence 1990 *Options for Change.* UK Government, London, retrieved on 2 September 2024, https://researchbriefings.files.parliament.uk/documents/CBP-7313/CBP-7313.pdf

———— 1998 *Strategic Defence Review*, UK Government, London, retrieved on 2 September 2024, https://researchbriefings.files.parliament.uk/documents/RP98-91/RP98-91.pdf

———— 2009 *The Strategic Review of Reserves*, UK Government, London, retrieved on 2 September 2024, https://assets.publishing.service.gov.uk/media/5a78da21ed915d0422065d95/strategic-defence-security-review.pdf

———— 2011 *Future Reserves 2020 – The Independent Commission to Review the United Kingdom's Reserve Forces*, UK Government, London, retrieved on 2 September 2024, https://www.gov.uk/government/publications/future-reserves-2020-study-fr20-final-report--2

——— 2013 *Reserves in the Future Force 2020, Valuable and Valued*, UK Government, London. retrieved on 2 September 2024, https://www.gov.uk/government/consultations/future-reserves-2020-consultation

——— 2014 *A History of our Reserves.* UK Government, London, retrieved on 2 September 2024, https://assets.publishing.service.gov.uk/media/5a7c054140f0b63f7572ad85/a-history-our-Reserves-Epub-v2.pdf

——— 2021 *UK Reserve Forces Review 2030.* UK Government, London, retrieved on 2 September 2024, https://www.gov.uk/government/publications/reserve-forces-review-2030

North Atlantic Treaty Organisation

North Atlantic Treaty Organisation 1949, NATO Treaty, NATO website, retrieved 13 January 2021, https://www.nato.int/cps/en/natolive/official_texts_17120.htm

Books

Allison, G 2017 *Destined for war: Can America and China escape Thucydides' trap?*, Houghton Mifflin Harcourt, Boston, MA

Ball, Desmond and Keiko Tamura (eds) 2013 Breaking Japanese Diplomatic Codes, David Sissons and D Special Section during the Second World War, ANU E-Press, Canberra https://press.anu.edu.au/publications/series/asian-studies/breaking-japanese-diplomatic-codes

Bekjr Ilhan, 2020 *China's Evolving Military Doctrine after the Cold War,* Siyaset, Ekonomi Ve Toplum Araştirmalari Vakfi Foundation for Political, Economic and Social Research, Ankora, Turkey

Blaxland J 2014 *The Australian Army from Whitlam to Howard*, Cambridge University Press, Melbourne

Blaxland, John and Clare Birgin, 2023 Revealing Secrets: An unofficial history of Australian signals intelligence and the advent of cyber, UNSW Press, Sydney, https://unsw.press/books/revealing-secrets/

Bleakley, Jack, 1992 The Eavesdroppers, Australian Government Publishing Service, Canberra, https://catalogue.nla.gov.au/catalog/252172

Bou, Jean., MacArthur's Secret Bureau: The story of the Central Bureau, General MacArthur's signals intelligence organisation, Australian Military History Publications, Loftus, NSW, 2012

Brady, AM 2003, *Making the foreign serve China: managing foreigners in the People's Republic*, Rowman & Littlefield, Maryland

Brady AM ed. 2012, *China's Thought Management*, Routledge, New York, retrieved 10 June 2020, https://trove.nla.gov.au/work/38962534?q&versionId=51751347

Blake, Greg 2019 Jungle Cavalry: Australian Independent Companies and Commandos 1941–1945, Helion and Company, Warwick, UK, 2019

Ballard, Geoffrey 1991 On Ultra Active Service, The Story of Australia's Signals Intelligence Operations during World War 2, Spectrum Publications, Richmond VIC, 1991

Breen Bob 2001 *Mission accomplished, East Timor, the Australian Defence Force participation in the International Force East Timor (INTERFET)*, Allen & Unwin, St Leonards

——— 2008 Struggling for Self-Reliance: Four case studies of Australian Regional Force Projection in the late 1980s and the 1990s. Australian National University Press, Canberra, retrieved 3 March 2022, https://press.anu.edu.au/publications/series/sdsc/struggling-self-reliance

——— 2016 *The Good Neighbour: Australian Peace Support Operations in the Pacific Islands 1980–2006*, Cambridge University Press, Melbourne https://www.cambridge.org/core/books/abs/good-neighbour/official-history-of-australian-peacekeeping-humanitarian-and-postcold-war-operations/D96C6C5F4F6BFC5B512E3812D7B2CCDC

——— 2022 *Disappointing the Dragon: how Australia should stand up to the Communist Party of China*, Echo Books, Melbourne, https://echobooks.com.au/our-books/disappointing-the-dragon/

Cheng D 2016, *Cyber Dragon: Inside China's Information Warfare and Cyber operations*, Praeger Security International: Santa Barbara

Clausewitz, C 1984, Howard, M & Paret, P eds., *On War*, Princeton University Press, United States

Cronin, PM ed. 2008, *The Impenetrable Fog of War: Reflections on Modern Warfare and Strategic Surprise*, Greenwood Press Group, Praeger Security International, Westport, CT

Davies, W 2021, *Special and Secret, The untold story of Z Special Unit in the Second World War*, Vintage Books, Australia

Dewar, M 1985, *British Army in Northern Ireland*, Guild Publishing, Wilts, UK

Doshi, R 2021, The long game: *China's grand strategy to displace American order*, Oxford University Press: New York

Fahey, John, 2023, The Factory, 2023, The official history of the Australian Signals Directorate, vol 1, Allen and Unwin, Sydney

Friedburg, AL 2011, *A Contest for Supremacy: China, America, and the Struggle for Mastery in Asia*, W.W. Norton and Company, New York

Jacques, M 2014, *When China Rules the World, When China Rules The World: The End of the Western World and the Birth of a New Global Order*, Penguin Books, London

Jian Hua To, J 2014, *Qiaowu: Extra-Territorial Policies for the Overseas Chinese*, Brill, Leiden, Netherlands, https://books.google.com.au/books?id=KGe7AwAAQBAJ&pg=PR5&dq=qiaowu%E2%80%99+strategy&source=gbs_selected_pages&cad=3#v=onepage&q=qiaowu%E2%80%99%20strategy&f=false

Jonsson, O 2019, *The Russian Understanding of War: blurring the lines between war and peace*, Georgetown University Press, Georgetown, Washington DC

Ganor, B 2015, *Global Alert: The Rationality of Modern Islamist Terrorism and the Challenge to the Liberal Democratic World*, Columbia University Press, New York

Grey J (2001) *The Australian Army, A History*, Oxford University Press, Melbourne

Hai-Chi Loo, J, Shiu Hing Lo, S and Chung-Fun Hung, S 2019, *China's New United Front Work in Hong Kong: Penetrative Politics and Its Implications*, Palgrave MacMillan, Singapore

Hamill, I 1981, *The Strategic Illusion: The Singapore Strategy for the defence of Australia and New Zealand, 1919–1942*, Singapore University Press, Singapore

Hamilton, C 2018, *Silent Invasion: China's influence in Australia*, Hardie Grant Books, Melbourne

Hamilton, C & Ohlbereg, M 2020, *Hidden Hand: Exposing how the Chinese Communist party is reshaping the world*, Hardie Grant Books, Sydney

Hoffman, F 2007, *Conflict in the 21st Century: The Rise of Hybrid War*, Potomac Institute for Policy Studies, Arlington, VA, retrieved 23 January 2020, https://www.potomacinstitute.org/images/stories/publications/potomac_hybridwar_0108.pdf

——— 2007, *Thoughts on 21st century warfare*, Potomac Press, Washington

Horner, D 1982, *High Command: Australia and Allied Strategy, 1939–1945*, Australian War Memorial and Allen and Unwin, Sydney

——— 1996, *Inside the War Cabinet: Directing Australia's War Effort 1939–1945*, Allen and Unwin, Sydney

——— 1998, *Blamey: Commander in Chief*, Allen and Unwin, Sydney

Ikenberry GJ 2015, *America, China, and the Struggle for World Order: Ideas, Traditions, Historical Legacies and Global Visions*, Palgrave, New York

Kilcullen, D 2020, *The Dragons and the Snakes: How the Rest Learned to Fight the West*, Scribe, Melbourne

Liang, Q & Xiangsui, W 1999 *Unrestricted Warfare: China's Master Plan to Destroy America*, 2017 Kindle Edition, Shadow Lawn Press, New York

Mazarr, MJ 2015, *Mastering the gray zone: understanding a changing era of conflict*, US Army War College Carlisle

McDonald, H, Ball, D, Dunn, J, van Kinken, G, Bourchier, D, Kammen, D & Tanter R 2002, *Masters of Terror, Indonesia's military and violence in East Timor in 1999*, Policy Paper no 145, Strategic and Defence Studies Centre, ANU, Canberra

Neville, L 2015, *Special Forces and the War on Terror*, Osprey Publishing, Oxford, UK

Palazzo A 2001 *The Australian Army – A History of its Organisation 1901–2001*, Oxford University Press, Melbourne

——— 2015 Forging Australian Land Power, A Primer, Australian Army Research Centre, Canberra

Petit, B 2013, *Going Big by Getting Small: The Application of Operational Art by Special Operations in Phase Zero*, Outskirts Press, Parker, CO

Pillsbury, M 2015, *The Hundred-Year Marathon: China's Secret Strategy to Replace America as the Global Superpower*, Henry Holt &Company, New York

Powell, Alan., 1996, War by Stealth: Australians and the Allied Intelligence Bureau, 1942–45, Melbourne University Press, Carlton, VIC

Raby, G 2020, China's Grand Strategy and Australia's Future in the New Global Order, Melbourne University Press, Melbourne

Rid, T 2013, *Cyber war will not take place*, Oxford University Press, UK

Rudd, Kevin, 2024, On Xi Jinping, How Xi's Marxist Nationalism is Shaping China and the World, Oxford University Press

Sanger, DE 2018, *The Perfect Weapon*, Broadway Books, New York

Schelling, TC 1966, *Arms and Influence*, Yale University Press, New Haven

——— 1980, *The Strategy of Conflict*, Harvard University Press, CT

Seidman, D 2011, *How: Why We Do Anything Means Everything*, John Wiley and Sons Publishing

Singer, PW and Brooking, ET 2018, *LikeWar: The Weaponization of Social Media*, Houghton Mifflin Harcourt Publishing Company, New York

Silver, L 2001, *Krait, the Fishing Boat that went to War*, Cultured Lotus, Sydney

Strachan, H 2018, *The Direction of War: Contemporary Strategy in Historical Perspective*, Cambridge University Press, Cambridge, UK

Urban, M 2012, *Big Boys' Rules: The SAS and the Secret Struggle Against the IRA*, Faber and Faber, London, UK

Valeriano, B and Maness, RC 2015, *Cyber War Versus Cyber Realities*, Oxford University Press, New York

White, H 2013, *The China Choice: why America should share power*, Black Inc, Melbourne

——— 2020, *How to Defend Australia*, Latrobe University Press/Black Inc, Melbourne

Yu-shek Cheng, J 2020, *Political Development in Hong Kong*, World Scientific Publishing, Singapore

Book Chapters/Theses

Armstrong, M 2024 Historical Legacy to National Asset: the Australian Army Reserve in the 21st Century, PhD Thesis, Deakin University, submitted June 2024

Breen, B 2006. *Australian Military Force Projection in the late 1980s and the 1990s: What happened and Why*, PhD Thesis, Australian National University, Canberra, retrieved 3 March 2022, https://openresearch-repository.anu.edu.au/bitstream/1885/7158/10/Breen-whole.pdf

——— 2016 'The strategic builder, creating an adaptive army for the future' in Connery, D. (ed) The Battles before, Case studies of Australian Army leadership after the Vietnam War, Army History Unit, Canberra

——— 2023 'The Kangaroo Exercise Series 1989–1995, Rehearsing mobilisation and force projection', Chapter 16 in Blaxland J and De Vogel M (eds), *Mobilising the Army, Contingencies and compromises over more than a century*, Proceedings of the 2021 Chief of Army's Conference, Canberra, 2023

Bury, R.G 1967, Plato, Plato in Twelve Volumes, vols 10 & 11, Harvard University Press, Cambridge, MA, retrieved 14 August 2020, http://www.perseus.tufts.edu/hopper/text?doc=Perseus%3Atext%3A1999.01.0166%3Abook%3D1%3Apage%3D626

Connolly P, 2023, 'Statecraft and Pushback: Delivering China's Grand Strategy in Melanesia 2014–2022', PhD Thesis, Australian National University, retrieved 23 June 2024, https://openresearch-repository.anu.edu.au/items/c5902078-7550-4d55-848e-26aa887a39e9

Costello J. and McReynolds, J 2020, 'China's Strategic Support Force: A Force for a New Era', Chapter 12 in Saunders, P.C, Ding, A.S., Scobell, A., Yang, A.N.D. and Wuthnow, J (eds) 2020, Chairman Xi Remakes the PLA: Assessing Chinese Military Reforms, National Defense University Press, Washington DC, retrieved 31 January 2020, https://ndupress.ndu.edu/Portals/68/Documents/Books/Chairman-Xi/Chairman-Xi.pdf

Dew, A 2008, 'The Erosion of Constraints in Armed-Group Warfare: Bloody Tactics and Vulnerable Targets' in Norwitz, J ed. 2008, *Armed Groups: Studies in National Security, Counterterrorism and Counterinsurgency*, US Naval War College, Newport, RI

Kouretsos, P 2019, 'A Literature Review' in Annex A: Contextualising Chinese Hybrid Warfare in Babbage, R 2019 *Stealing a March: Chinese Hybrid Warfare in the Indo-Pacific: Issues and Options for Allied Defense Planners, vol. II Case Studies*, Center for Strategic and Budgetary Assessments, Washington DC, retrieved 12 February 2020, https://csbaonline.org/research/publications/stealing-a-march-chinese-hybrid-warfare-in-the-indo-pacific-issues-and-options-for-allied-defense-planners

Langford, I.D 2023, The Grey Zone: When war is not war a de-escalation strategy for Australia, Deakin University, https://dro.deakin.edu.au/articles/thesis/The_Grey_Zone_When_war_is_not_war_a_de-escalation_strategy_for_Australia/22735115

Norberg, J, Westerlund, F & Franke, U 2014, 'The Crimea operation. Implications for future Russian military interventions' in Granholm, N, Malminen, J & Persson, G eds. 2014, *A rude awakening. ramifications of Russian aggression towards Ukraine*, FOI, Stockholm

Shemella, P 2011, *Fighting Back: What Governments Can Do About Terrorism*, 1st edn., Stanford University Press, Redwood City, CA

Sparrow, R 2011, 'Robotic Weapons and the Future of War', pp. 117–33, in Tripodi, P & Wolfendale J eds., *New Wars and New Soldiers: Military Ethics in the Contemporary World*, Ashgate Publishing Ltd, Surrey, UK

Symon, P 2023, 'Yes, Australia does need a national security advisor,' 20 December, Australian Security Policy institute, *The Strategist*, retrieved 24 June 2024, https://www.aspistrategist.org.au/yes-australia-does-need-a-national-security-adviser/

Taft, E 2017, 'Outer Space: The Final Frontier or the Final Battlefield?', vol. 15, no. 1, *Duke Law & Technology Review*, Duke University, Durham, NC

Whitmore, P 2016 'Current Cyber Wars', in Janczewski, LJ and Caelli, W 2016, Cyber Conflicts and Small States, Ashgate Publishing, Dorchester

Monographs

Babbage, R 2015, *Game Plan: The Case for a New Australian Grand Strategy*, R.G. Menzies Essay, Menzies Research Centre, Connor Court Publishers, Ballarat

——— 2016, *Countering Chinese Adventurism in the South-China Sea,* Centre for Budgetary and Strategic Assessments, Washington, DC

——— 2019, *Stealing a March: Chinese Hybrid Warfare in the Indo-Pacific; Issues and Options for Allied Defense Planners*, Centre for Budgetary and Strategic Assessments, Washington, DC

——— 2019, *Winning Without Fighting: Chinese and Russian Political Warfare Campaigns and how the West can prevail*, Vol 1 and 2, Centre for Budgetary and Strategic Assessments, Washington, DC

——— 2020, *Which Way the Dragon: sharpening allied perceptions of China's strategic trajectory*, Center for Strategic and Budgetary Assessments, Washington DC

——— Mahnken, T and Toshihara, T 2018*Countering Comprehensive Coercion-Competitive strategies against authoritarian political warfare,* Centre for Budgetary and Strategic Assessments, Washington, DC

Bassi, Justin, Foreword in Bec Shrimpton et al, The cost of Defence: ASPI Defence budget brief 2024–2025, Australian Strategic Policy Institute, Canberra, June 2024

Breen B 2014 *Preparing the Australian Army for joint employment: a short history of the Adaptive Army Initiative 2007–2010*, Australian Army, Canberra

Blaxland, J 2019, 'A Geostrategic SWOT Analysis for Australia', Centre of Gravity series, no. 49, Strategic and Defence Studies Centre, ANU, Canberra June, retrieved 3 July 2020. https://sdsc.bellschool.anu.edu.au/sites/default/files/publications/attachments/2019-06/cog_49_swot_analysis_web.pdf

Burke, E.J., Gunness, K, Cooper C.A. Cozad M.I, 2020, *People's Liberation Army Operational Concepts,* Rand Report, Rand Corporation, retrieved 31 January 2020, https://www.rand.org/pubs/research_reports/RRA394-1.html

Fox, AC 2017, *Hybrid Warfare: The 21st Century Russian Way of War*, School of Advanced Military Studies, Defense Technical Information Center Leavenworth, KS, retrieved 4 July 2020, https://apps.dtic.mil/sti/citations/AD1038987

Hartcher, P 2019, 'Red Flag: Waking up to China's challenge', *Quarterly Essay,* Black Inc Publishing, Sydney

Knight, C & Theodorakis, K 2019, *The Malawi crisis – urban conflict and information operations*, Special Report, Australian Strategic Policy Institute, Canberra

Kofman, M, Migacheva, K, Nichiporuk, B, Radin, A, Tkacheva, O & Oberholtzer J 2017, *Lessons from Russia's Operations in Crimea and Eastern Ukraine*, The Rand Corporation, retrieved 15 June 2020, https://www.rand.org/pubs/research_reports/RR1498.html

Langford, I 2014, Australian special operations: Principles and considerations, Australian Army, Department of Defence, Canberra

Morris, LJ, Mazarr, MJ, Hornung, JW, Pezard, S, Binnendijk, A and Keep, M 2019, Gaining Competitive Advantage in the Gray Zone. Response Options for Coercive Aggression Below the Threshold of Major War, Rand Corporation, CA, retrieved 19 February 2021, https://www.rand.org/pubs/research_reports/RR2942.html

Norberg, J 2016, 'The use of Russia's military in the Crimea crisis.' Carnegie Endowment for International Peace Russia Programme, retrieved 15 February 2018, http://carnegieendowment.org/2014/03/13/use-of-russia-s-military-in-crimean-crisis

Parker, S & Chetfitz, G 2018, *Debtbook Diplomacy: China's Strategic Leveraging of its Newfound Economic Influence and the Consequences for U.S. Foreign Policy*, Belfer Center for Science and International Affairs, Harvard Kennedy School, Cambridge, MA, retrieved 20 June 2020, https://www.belfercenter.org/sites/default/files/files/publication/Debtbook%20Diplomacy%20PDF.pdf

Paul, C and Matthews, M 2016, *The Russian "Firehose of Falsehood" Propaganda Model,* RAND Corporation Santa Monica, California, retrieved 23 June 2024, https://www.rand.org/pubs/perspectives/PE198.html

White, H 2017, *Without America: Australia in the New Asia*, The Quarterly Essay, Black Inc, Melbourne

Papers/Journals

Lowy Institute 2020 '2020 Lowy Institute Opinion Poll,' June, Lowy Institute, retrieved 2 July 2020, https://poll.lowyinstitute.org/themes/security-and-defence

Azad TM, Haider MW and Sadiq M 2022 'Understanding Grey Zone Warfare from multiple perspectives', *World Affairs*, 186:1, pp. 81–104

Banks G 2010 Advancing Australia's 'Human Capital Agenda', Presentation at the Ian Little Lecture, Productivity Commission, 13 April 2010, https://www.pc.gov.au/media-speeches/speeches/advancing-human-capital (retrieved 02 August 2023)

Bergin A and Templeman D 2020 'National Bushfires Emergency? Let's Have a National Response', Australian Strategic Policy Institute, 6 January

Blanchette J (2021) 'Xi's gamble: the race to consolidate power and stave off disaster', *Foreign Affairs*, July/August 2021, Vol. 100, No. 4., pp. 10–19

Bland D and Shimooka R (2011) *Let Sleeping Dogs Lie, The Influence of External Studies and Reports on National Defence Policy – 2000 to 2006*, School of Policy Studies, Queen's University, Kingston

Bachmann, S, Oliver, V, Dowse, A & Gunneriusson, H 2019, 'Competition Short of War – How Russia's Hybrid and Grey-Zone Warfare are a Blueprint for China's Global Power Ambitions', *Australian Journal of Defence and Strategic Studies*, vol. 1, no.1, retrieved 13 March 2020, https://papers.ssrn.com/sol3/papers.cfm?abstract_id=3483981

Bendett, S and Kania, E 2019, A New Sino-Russian High Tech Partnership: Authoritarian innovation in an era of great-power rivalry, Policy Brief, Report No. 22 2019, Australian Strategic Policy Institute – International Cyber Policy Centre, 29 October, Canberra, retrieved 24 December 2020, https://s3-ap-southeast-2.amazonaws.com/ad-aspi/2019-10/A%20new%20Sino-Russian%20high-tech%20partnership_0.pdf?xAs9Tv5F.GwoKPiV9QpQ4H8uCOet6Lvh

Blaxland J 2019 *A Geostrategic SWOT Analysis for Australia*, The Centre of Gravity Series Australian National University, Canberra

Blaxland, J 2020, *Developing a new Plan B for the ADF: Implications from a Geostrategic SWOT Analysis for Australia*, Australian Army Research Centre, retrieved 3 July 2020, https://researchcentre.army.gov.au/library/seminar-series/developing-plan-b-adf-some-implications-arising-geostrategic-swot-analysis-australia

Bowman S, Kapp L and Belasco A 2005 *Hurricane Katrina, DOD Disaster Response* (Washington, Congressional Research Service)

Bongiorno, F 2017, 'Up to a point, Professor Hamilton', *Inside Story*, Inside Story Publishing, Melbourne, retrieved 13 June 2020, http://insidestory.org.au/up-to-a-point-professor-hamilton/

Brady AM 2017, 'Magic Weapons: China's political influence activities under Xi Jinping', *Wilson Center*, 18 November, retrieved 2 June 2020, https://www.wilsoncenter.org/article/magic-weapons-chinas-political-influence-activities-under-xi-jinping

Brady, AM ed. 2010, 'Looking North, Looking South: China, Taiwan, and the South Pacific', *World Scientific*, vol. 26, series on contemporary China, https://www.bookdepository.com/Looking-North-Looking-South-China-Taiwan-South-Pacific-Anne-Marie-Brady/9789814304382

Brady S, Hanasz P and Bigland P 2019 Aiding the civil authority, the potential for a broader Army role in domestic counter-terrorism, Australian Army Occasional Paper Developmental Series, No. 2, Australian Army Research Centre, https://researchcentre.army.gov.au/library/occasional-papers/aiding-civil-authority-potential-broader-army-role-domestic-counter-terrorism (retrieved 19 July 2023)

Brangwin N, Church N, Dyer S and Watt D 2015 Defending Australia, a history of Australia's Defence White Papers'. Parliamentary Library Research Paper. Department of Parliamentary Services, Canberra

Brangwin, N 2020 *Australia's military involvement in Afghanistan since 2001: a chronology*, Parliament of Australia, Parliament Library, Canberra

Bree M (2020) 'Answering trunk call with some inside knowledge', *Defence News*, 15 January 2020, retrieved 25 February 2020, https://news.defence.gov.au/national/answering-trunk-call-some-inside-knowledge

Cantwell D (2017) Hybrid Warfare: Aggression and Coercion in the Grey Zone, *American Society of International Law ASIL Insights*, 29 November, Vol. 21, Issue: 14

Carment, D & Belo D 2018, 'War's Future: The Risks and Rewards of Grey Zone Conflict and Hybrid Warfare', *Policy Paper*, Canadian Global Affairs Institute

Cavallaro G 2017 Army National Guard 4.0, Effort means 'evolutionary leap' for citizen soldiers, Association of the US Army, 03 October, retrieved 20 April 2020, https://www.ausa.org/articles/army-national-guard-evolutionary-leap-citizen-soldiers

Center for Security Policy 2023 War in Ukraine: Lessons Identified and Learned, European Values, retrieved 24 July 2023, https://europeanvalues.cz/en/war-in-ukraine-lessons-identified-and-learned/

Church N 2014 The evolution of the Australian Defence Force Gap Year program, Research paper series 203–14, 08 May 2014, Parliamentary Library: Canberra

CIA (Central Intelligence Agency) (2023). *World Factbook*, retrieved 29 July 2023 www.cia.gov/library/publications/the-world-factbook

Clay P 2013 The Australian Army's 2nd Division, an update, address to the NSW Royal United Service Institute, 24 September 2013, retrieved 06 April 2023, https://www.rusinsw.org.au/Papers/20130924.pdf

Coates J and Smith K 1995 *Review of the Ready Reserve Scheme*. 30 June. UNSW, Canberra

Combined Arms Training Centre 2013 *Conduct Reserve Response Force (RRF) Operations Training Management Package*, Version 4.1, 15 October 2013, Australian Army, Canberra

Correll JT 2011 'Origins of the Total Force.' *Air Force Magazine,* Vol. 94, No. 2 February 2011

Council of Reserve Forces' and Cadets' Associations 2019 *The United Kingdom Reserve Forces External Scrutiny Team annual report 2019*, The Council of RFCAs, London

Crompvoets S 2014 Exploring future service needs of Australian Defence Force Reservists, Australian National University, Canberra

Crompvoets S 2021 'Blood, Lust Trust and Blame', Monash University Publishing, Melbourne

Cull M (2020) 'Value Beyond Money, Australia's Special Dependence on Volunteer Firefighters', *The Conversation*, 23 January, retrieved 18 March 2020, http://theconversation.com/value-beyond-money-australias-special-dependence-on-volunteer-firefighters-129881

Cossacks, C, Ottis, R & Taliham, AM 2011, Estonia after the 2007 Cyberattacks: legal, strategic and organisational changes in cyber security, case studies in information warfare and security, Cooperative Cyber Defence Centre of Excellence, Estonia, International Journal of Cyber Warfare and Terrorism, retrieved 13 January 2021, https://www.igi-global.com/article/estonia-after-2007-cyber-attacks/61328

Davies A and Smith H 2008 'Stepping Up, Parttime Forces and ADF capability', *Strategic Insights,* Australian Strategic Policy Institute, Canberra

Davies, A, Jennings, P and Schreer, B 2014, A versatile force- the future of Australia's Special Forces, Australian Strategic Policy Institute, Canberra

Davis A 2019 'A new DWP wouldn't be worth the white paper it's written on'. *The Strategist.* https://www.aspistrategist.org.au/a-new-dwp-wouldnt-be-worth-the-white-paper-its-written-on/

Defence Reserve Association (DRA) 2016, Submission to the Review of the Defence Annual Report 2015–16, retrieved 05 April 2023, https://www.aph.gov.au/DocumentStore.ashx?id=6f6b7837-7ea5-4f42-a8d6-3d4368a3744c&subId=463019

Defence Reserves Support Council 2020 'Call-out Orders', retrieved 14 February 2020, https://www.defencereservessupport.gov.au/call-out-orders/

Defence Reserves Support Council (2020b) 'Call-out Orders – Information for Reservists', at https://www.defencereservessupport.gov.au/call-out-orders/information-for-reservists/ (retrieved 20 February 2020)

Defence Statistics 2013 TSP7-UK Reserve Forces and Cadets, 01 April 2013, Ministry of Defence, London

Dibb, P 2019, 'How the geopolitical partnership between China and Russia threatens the West', Special Report, SR148, 29 November, *Australian Strategic Policy Institute*, Canberra, retrieved 2 May 2020, https://www.aspi.org.au/report/how-geopolitical-partnership-between-china-and-russia-threatens-west

Dibb, P, Brabin-Smith R & Sargeant B 2018, 'Why Australia Needs a Radically New Defence Policy', The Centre of Gravity series, Strategic and Defence Studies Centre, ANU, Canberra

Fawcett D (2015) 'Speech by Senator David Fawcett to the Defence Reserve Association National Conference', 22 August 2015, retrieved 20 Oct 2023, https://dra.org.au/conference-2015-item/14921/transcript-of-speech-by-senator-david-fawcett-to-the-dra-national-conference/?type_fr=346

Feeney D 2011 'ADF Reserve Capability, Where to now?', Presentation to Defence Reserves Association Conference, 20 August 2011, retrieved 05 April 2023 https://parlinfo.aph.gov.au/parlInfo/search/search.w3p

Finney N and Mayfield T (eds) 2018 *Redefining the Modern Military: The Intersection of Profession and Ethics*, Naval Institute Press, Annapolis

Ergas H and Thomson M 2011 'More Guns Without Less Butter, Improving Australian Defence Efficiency', *ANU Agenda*, Vol 18, No 3 2011 pp. 31–62

Gerasimov, V, 2013, 'The Value of Science is in the Foresight: New Challenges Demand Rethinking the Forms and Methods of Carrying out Combat Operations,' Voyenno-Promyshlennyy Kurier, 26 February

Gruen D 2000 *The Australian Economy in the 1990s*, Reserve Bank of Australia Bulletin, Oct 2000

Hamilton, C 2018 Why do we keep turning a blind eye to Chinese political interference? 4 April, *The Conversation*, retrieved 19 April 2020, https://theconversation.com/why-do-we-keep-turning-a-blind-eye-to-chinese-political-interference-94299

Hawke A and Smith R 2012 *Australian Defence Force Posture Review* Canberra, Department of Defence

Hayward-Jones, J 2013, 'Big Enough for All of Us: Geo-strategic Competition in the Pacific Islands', *Commentary*, Lowy Institute, May 2013

Hayward-Jones, J 2014, 'Australia's costly investment in Solomon Islands: The Lessons of RAMSI', The Lowy Institute, 8 May, retrieved 3 May 2021, https://www.lowyinstitute.org/publications/australias-costly-investment-solomon-islands-lessons-ramsi

Helmly JR 2004 *Readiness of the Army Reserve*, Memorandum dated 20 December 2004, https://www.globalsecurity.org/military/library/report/2005/usar_memo-20dec2004.htm (retrieved 10 May 2023)

Hellyer, M 2020, The Cost of Defence 2020–2021Part 1: ASPI 2020–2021 Defence Budget Brief, 12 August, The Cost of Defence 2020–2021Part 2: ASPI 2020–2021 Defence Budget Brief, 22 October, Australian Strategic Policy Institute, retrieved 28 April 2021, https://www.aspi.org.au/report/cost-defence-2020-2021-part-1-aspi-2020-strategic-update-brief

Hunter F, Impiombato D, Lau Y and Trigs A 2023 Countering China's Coercive Diplomacy Prioritising economic security, sovereignty and the rules-based order, *Policy Brief Report No. 68/2023*, ASPI: Canberra

Hunter P 2006 'Stakeholder Perspectives on the Special Commission on the restructuring of the Reserves, 10 years later', *Journal of Military and Strategic Studies*, Winter 2005–2006, Vol. 8, Issue 2

Irwin LG 2019 *A Modern Army Reserve for a Multi-Domain World, Structural Realities and Untapped Potential* (Carlisle, US Army War College Press)

Jakobson, L 2021, Why should Australia be concerned about rising tensions in the Taiwan Straits?, China Matters Explores, 9 February, retrieved 9 February 2021, https://chinamatters.org.au/policy-brief/policy-brief-february-2021/

Jans, N and Schmidchen, D 2003 'Culture and organisational behaviour at Australian Defence Headquarters', *Australian Defence Force Journal*, No. 158, January/February, pp.23–28

Jennings P 2016 'The Politics and Practicalities of Designing Australia's Force Structure', in Ball, D. and Sheryn Lee S. (ed) 2016, *Geography, Power, Strategy and Defence Policy, Essays in Honour of Paul Dibb*, ANU Press, Canberra

Jiji Press 2019 'Japan Calls Up SDF Reserves for First Time Since 3/11 for Typhoon Hagibis Disaster Relief', *Japan Times*, 28 October

Johnston M 2013 *The Australian Army in World War II.* Bloomsbury Publishing, London

Joske, A 2020, 'The Party Speaks for You: Foreign interference and the Chinese Communist Party's United Front System', *Policy Brief*, Report no. 32/2020, 9 June, Australian Strategic Policy Institute, Canberra, retrieved 10 June 2020, https://www.aspi.org.au/report/party-speaks-you

Kania, E 2020, Innovation in the New Era of Chinese Military Power: What to make of the new Chinese defense white paper, the first since 2015, *The Diplomat*, retrieved 31 January 2020 https://thediplomat.com/2019/07/innovation-in-the-new-era-of-chinese-military-power/

Kennan, GF 1948, 'The Inauguration of Organized Political Warfare [redacted version]', Wilson Center, Digital Archive, International History Declassified, retrieved 6 August 2020 at http://www.digitalarchive.wilsoncenter.org

Kidson R (2017) *Force Design in the 1990s, Lessons for contemporary change management, A retrospective appraisal of Army in the 21st Century and Restructuring the Army*, an Australian Army History monograph, Canberra, 2017, https://researchcentre.army.gov.au/library/occasional-papers/force-design-1990s retrieved 27 May 2023

Kidson R 2020 '5th Engineer Regiment', *Australian Sapper 2020*, Magazine of the Royal Australian Engineers, https://raevictoria.files.wordpress.com/2021/01/7._australian_sapper_2020_eversion.pdf (retrieved 02 August 2023)

——— 2022 *Scaling the Force, Reserve Mobilisation for Domestic Contingencies, Working Paper*, Australian Army History Unit, Canberra

Kilsby A 2023 History of the Defense Reserve Association, presentation to the 2023 Defence Reserve Association, 12 August 2023

Kofman, M & Rojansky, M 2015, 'A closer look at Russia's 'Hybrid War'', *Kennan Cable*, no. 7, Wilson Center/Kennan Center, Washington DC, retrieved 23 May 2020 https://www.files.ethz.ch/isn/190090/5-KENNAN%20CABLE-ROJANSKY%20KOFMAN.pdf

Kramer M 2014 Sequestration's Impact on Military Spending 2013 – 2014, National Priorities Project retrieved 14 April 2019 https://www.nationalpriorities.org/analysis/2014/sequestration-impact-on-military-spending-2013-2014/

Leahy P 2004 'Towards the hardened and networked army'. *Australian Army Journal*, 2(1), 27–36

——— 2007 'From the Source, LTGEN Peter Leahy AC Chief of Army, *Australian Defence Magazine*, October 2007, retrieved 19 July 2007, https://www.australiandefence.com.au/

Maclellan, S and O'Leary, N 2017, 'Doing Battle in Cyberspace: How an Attack on Estonia Changed the Rules of the Game', Centre for International Governance Innovation, Cybersecurity, Surveillance and Privacy, retrieved 20 October 2020, https://www.cigionline.org/

Norberg, J 2016, 'The use of Russia's military in the Crimea crisis.' Carnegie Endowment for International Peace Russia Programme, retrieved 15 February 2018, https://carnegieendowment.org/2014/03/13/use-of-russia-s-military-in-crimean-crisis

Nye, JS 2010, 'Cyber Power', Essay from the Belfer Centre for Science and International Affairs, Harvard Kennedy School, May

O'Neill, M 2020, 'Punching at Air: The military and the Grey Zone', 26 June, *Land Power Forum*, Australian Army Research Centre, Canberra, retrieved 13 July 2020, https://researchcentre.army.gov.au/library/land-power-forum/punching-air-military-and-grey-zone

Smith, P 2020, *Russian Electronic Warfare: A Growing Threat to U.S. Battlefield Supremacy*, American Security project, retrieved 20 October 2020 at http://www.americansecurityproject.org

Varghese, P 2019, 'A new China narrative for Australia Submission by Peter Varghese,' *China Matters*, 23 April 2019, retrieved 6 July 2020, http://chinamatters.org.au/wp-content/uploads/2019/04/China-Matters_A-new-China-Narrative-Submission_Peter-Varghese-23042019.pdf

Varghese, P 2019, What should Australia do to manage risk in its relationship with the PRC? *China Matters*, retrieved 6 July 2020, http://chinamatters.org.au/policy-brief/policy-brief-june-2020/

Wallis, J 2015, 'The South Pacific: arch of instability of arch of opportunity?', vol. 27, Iss. 1: The Sixth Oceanic Conference on International Studies: Transitions in the Asia Pacific, *Global Change, Peace & Security*, 3 February, pp. 39–53, retrieved 20 June 2020, https://www.tandfonline.com/doi/abs/10.1080/14781158.2015.992010?journalCode=cpar20

White, J 2016, 'Dismiss, Distort, Distract, and Dismay: Continuity and Change in Russian Disinformation,' Policy Brief 13, *Institute for European Studies*, May, retrieved 6 August 2020, http://www.ies.be/policy-brief/dismiss-distort-distract-and-dismay-continuity-and-change-russian-disinformation

Journal Articles

Ackerman, RK 2017, 'Russian Electronic Warfare Targets NATO Assets', *Signal magazine,* 01 November, retrieved 20 October 2020 http://www.afcea.org

Armstrong M 2011 'Not hearts and minds, civil-military cooperation in OBG (W)-3', *Australian Army Journal*, Vol. 8, No. 1, Autumn 2011

——— 2011 *Deployed interagency operations: operationalising the comprehensive approach*, Peace Operations Training Institute, New York. Retrieved on 23 June 2024, http://cdn.peaceopstraining.org/theses/armstrong.pdf

Ayson, R 2007 'The 'arch of instability' and Australia's strategic policy', vol. 61, iss. 2, 22 May, *Australian Journal of International Affairs*, pp. 215–231, retrieved 20 June 2020, https://www.tandfonline.com/doi/full/10.1080/10357710701358360?src=recsys

Barber, N 2017 'A warning from the Crimea: hybrid warfare and the challenge for the ADF', no. 201, *Australian Defence Force Journal*, Canberra, pp. 46–58, retrieved 4 July 2020, https://www.defence.gov.au/adc/adfj/Documents/issue_201/Barber_April_2017.pdf

Bendett, S and Kania, E 2020 'The Resilience of Sino-Russian High-Tech Cooperation', *War on the Rocks*, 5 August, retrieved 24 December 2020, https://warontherocks.com/2020/08/the-resilience-of-sino-russian-high-tech-cooperation/

Blasko, D 2015 'Chinese Special Forces: Not like 'Back at Bragg', *War on the Rocks*, retrieved 28 Jan 2019, https://warontherocks.com/2015/01/chinese-special-operations-forces-not-like-back-at-bragg/

Byman, D & Merritt, I 2018 'The New American Way of War: Special Operations Forces in the War on Terrorism', *The Washington Quarterly*, vol. 341, iss 2, Georgetown University, Elliott School of International Affairs, Washington DC

Carrico, K 2019, 'In defence of Silent Invasion', *Policy Forum*, Asia, the Pacific Policy Society, 1 March, retrieved 13 June 2020, https://www.policyforum.net/in-defence-of-silent-invasion/

Chansoria, M 2012, 'Defying Borders in Future Conflict in East Asia: Chinese Capabilities in the Realm of Information Warfare and Cyber Space', *Journal of East Asia Affairs*, Vol. 26, No. 1, pp. 105–06, retrieved 12 May 2020, https://www.jstor.org/stable/23257910?seq=1

Fitzgerald, J 2016, 'Beijing's *quoqing* versus Australia's way of life', 27 September, Inside Story, Inside Story Publishing, Melbourne, retrieved 8 June 2020, https://insidestory.org.au/beijings-guoqing-versus-australias-way-of-life/

Florcruz, M 2014 'Chinese Military Professor: Maritime Disputes will lead to WWIII', *Business Insider*, 18 September, retrieved 1 February 2017, https://www.businessinsider.com/michelle-florcruz-maritime-disputes-will-lead-to-world-war-iii-2014-9?IR=T

Glenn, RW 2009 'Thoughts on 'Hybrid Conflict'', *Small Wars Journal*, Bethesda, MD, retrieved 20 February 2020, https://smallwarsjournal.com/blog/journal/docs-temp/188-glenn.pdf

Gray, C.S 2010 'Gaining Compliance: The Theory of Deterrence and its Modern Application', vol. 29, iss 3, *Comparative Strategy*, 22 July, pp. 278–283, retrieved 14 February 2021, https://www.tandfonline.com/doi/full/10.1080/01495933.2010.492198?scroll=top&needAccess=true

Hollis, D 2011, 'Cyberwar Case Study: Georgia 2008', *smallwarsjournal.com*, retrieved 12 January 2022, https://smallwarsjournal.com/blog/journal/docs-temp/639-hollis.pdf

Huth, P and Russett, B 1984, 'What Makes Deterrence Work? Cases from 1900 to 1980', vol 36, no 4, *World Politics*, Cambridge University Press

Jakobson, L et al, 2021, A New China Narrative for Australia, *China Matters*, retrieved 9 February 2021, http://chinamatters.org.au/a-new-china-narrative-for-australia/text-of-a-new-china-narrative-for-australia/

Johnston, R 2018, 'Hybrid War and Its Countermeasures: A Critique of the Literature', *Small Wars and Insurgencies,* vol. 29, iss. 1, Routledge, Taylor & Francis Online, New York, pp. 141–163, retrieved 1 February 2020, https://www.tandfonline.com/doi/full/10.1080/09592318.2018.1404770?src=recsys

Karber, P and Thibeault, J 2016, 'Russia's New-Generation Warfare', *Association of the United States Army,* retrieved 14 August 2020 at http://www.ausa.org

Kofman, M 2020, 'The Emperor's League: Understanding Sino-Russian Defense Cooperation' *War on the Rocks,* 6 August, retrieved 24 December 2020, https://warontherocks.com/2020/08/the-emperors-league-understanding-sino-russian-defense-cooperation/

Langford, I 2017, 'Australia's Offset and A2/AD Strategies', no. 47, iss. 1, *Parameters,* Spring 2017, retrieved 10 November 2019, http://ssi.armywarcollege.edu/pubs/parameters/issues/Spring_2017/11_Langford_AustraliasOffsetAndA2AD.pdf

Nye JS 2017, 'Deterrence and Dissuasion in Cyberspace', *International Security,* Vol. 41, No. 3

Pearlman, J, Parfitt S & Parfitt, T 2014, 'Russia has sent a convoy of warships to Australia northern maritime border', *Business Insider,* 14 November, retrieved 2 June 2020, https://www.businessinsider.com/russian-warships-at-australias-border-2014-11?IR=T

Reilly, J 2018, 'The Multi-Domain Operations Strategist', *Over the Horizon Journal,* 8 November 2018, retrieved 20 May 2020 at https://othjournal.com/2018/11/08/oth-mdos-reilly/

Renz, B, 2016, 'Russia and 'hybrid war', *Contemporary Politics,* vol. 22, Iss. 3 in Russia the West and the Ukraine Crisis, pp. 283–300, retrieved 01 February 2020, https://www.tandfonline.com/doi/full/10.1080/13569775.2016.1201316?src=recsys

Schake, K 2019 'Social Media as War?', War on the Rocks, 5 September 2019, retrieved 5 May 2020 at https://warontherocks.com/2018/09/social-mdeia-as-war/

Segal, A 2020, Peering into the Future of Sino-Russian Cyber Security Cooperation, *War on the Rocks,* 10 August, retrieved 24 December 2020, https://warontherocks.com/2020/08/peering-into-the-future-of-sino-russian-cyber-security-cooperation/

Sechser, TS 2011, 'Militarized Compellent Threats, 1918–2001', *Conflict Management and Peace Science,* Peace Science Society (International), Sage Publications, Newbury Park, CA, retrieved on 16 January 2020, https://journals.sagepub.com/doi/abs/10.1177/0738894211413066

Stoker, D & Whiteside, C 2020, 'Blurred Lines: Gray-Zone Conflict and Hybrid War – Two Failures of American Strategic Thinking', vol 73, no 1, *Naval War College Review,* Newport RI, retrieved 3 February 2021 https://digital-commons.usnwc.edu/nwc-review/vol73/iss1/1

Strawser, BJ 2010, 'Moral Predators: The Duty to Employ Uninhabited Aerial Vehicles', vol. 9, iss. 4, *Journal of Military Ethics*, pp. 342–68, retrieved 28 May 2020, https://www.tandfonline.com/doi/abs/10.1080/15027570.2010.536403

Talesco, C 2020, 'Foreign Aid to Timor-Leste and the Rise of China', *Journal of International Studies*, pp. 131–150, retrieved 4 May 2020, https://www.jois.eu/

Thornton, R, 2015, 'The Changing Nature of Modern Warfare, Responding to Russian Information Warfare', *The RUSI Journal*, Vol 160, 2015 – Issue 4

Trevithick, J 2019, 'Ukrainian Officer Details Russian Electronic Warfare Tactics', *The War Zone,* retrieved 11 October 2020 at http://www.thedrive.com

Walker, C & Ludwig, J 2017 'The Meaning of Sharp Power', in *Foreign Affairs*, retrieved 10 November 2018, https://www.foreignaffairs.com/articles/china/2017-11-16/meaning-sharp-power

Watts, BD 2004 'Clauswitzian Friction and Future War', *McNair Papers*, Institute for National Strategic Studies-National Defence University, Washington, DC

Williams, G 2011 'A Decade of Australian Anti-Terror Laws', *Melbourne University Law Review*, vol. 35, iss. 3, retrieved 9 April 2020, http://classic.austlii.edu.au/au/journals/MelbULawRw/2011/38.html#Heading82

Zhang D & Lawson, S 2017 'China in Pacific Region Politics', The Round Table, *The Commonwealth Journal of International Affairs*, vol. 106, Iss 2: Pacific Region Politics, 18 April, pp. 197–206, retrieved 20 June 2020, https://www.tandfonline.com/doi/full/10.1080/00358533.2017.1296705?src=recsys

Transcripts

Chivvis, C 2017 'Understanding Russian 'Hybrid Warfare' and What Can Be Done about it', Testimony presented before the House Armed Services Committee on March 22', Rand Corporation, retrieved 07 February 2020, https://www.rand.org/pubs/testimonies/CT468.html

Grattan M 2014 'In conversation with ASIO chief David Irvine', *The Conversation*, August 15, retrieved 25 July 2019, https://theconversation.com/grattan-on-friday-in-conversation-with-asio-chief-david-irvine-30536

Irvine, D 2013 'Director-General Speech: Address to the Security in Government Conference, 2013- Australia's current security and intelligence operating environment', Attorney-General's Department, 13 August, Canberra

Lewis, D 2019 'Address to the Lowy Institute', 5 September, retrieved 26 April 2020, https://www.lowyinstitute.org/news-and-media/multimedia/audio/address-asio-director-general-duncan-lewis

Keating, P 2019 'Paul Keating's speech on Australia's China policy – full text,' *The Guardian*, 18 November, retrieved 13 June 2020, https://www.theguardian.com/australia-news/2019/nov/18/paul-keatings-speech-on-australias-china-policy-full-text

Mattis, P 2018 'Testimony before the U.S.-China Economic and Security Review Commission: Hearing on China's Relations with U.S. Allies and Partners in Europe and the Asia-Pacific', 5 April, U.S.-China Economic and Security Review Commission website, retrieved 17 February 2021, https://www.uscc.gov/sites/default/files/ USCC%20Hearing%20_Peter%20Mattis_Written%20Statement_April%205%20 2018.pdf

Morrison, S 2020 'Address – Launch of the 2020 Defence Strategic Update', 1 July, retrieved 13 July 20202, https://www.pm.gov.au/media/address-launch-2020-defence-strategic-update

Morrison, S 2019 'Stepping Up Australia's Response Against Foreign Interference', *media release*, Prime Minister of Australia website, 2 December, retrieved 19 July 2020, https://www.pm.gov.au/media/stepping-australias-response-against-foreign-interference

Morrison, S, Reynolds, L & Dutton, P 2020 'Statement on malicious cyber activity against Australian networks', *media release*, 19 June, Prime Minister of Australia website, retrieved 23 June 2020, https://www.pm.gov.au/media/statement-malicious-cyber-activity-against-australian-networks

Morrison, S 2020 'Australia's largest ever investment in cyber security', *media release*, 30 June, Prime Minister of Australia website, retrieved 1 July 2020, https://www. pm.gov.au/media/nations-largest-ever-investment-cyber-security

Nye, J 2011 The Future of Power, Transcript, Chatham House, London

Ottis, R 2008 'Analysis of the 2007 Cyber Attacks against Estonia from the Information Warfare Perspective' in Proceedings of the 7th European Conference on Information Warfare and Security, 30 June – 1 July 2008, Plymouth University, Academic Publishing Limited, Reading, pp 163–168, retrieved 17 December 2020, https://ccdcoe.org/library/publications/analysis-of-the-2007-cyber-attacks-against-estonia-from-the-information-warfare-perspective/

Payne, M 2021 'Joint statement on arrests in Hong Kong', DFAT website, 10 January, retrieved 17 February 2021, https://www.foreignminister.gov.au/minister/ marise-payne/media-release/joint-statement-arrests-hong-kong

Singapore Government, 'Transcript of a talk given by the Prime Minister, Mr. Lee Kuan Yew, on the subject "Big and Small Fishes in Asian Waters" at a meeting of the University of Singapore Democratic Socialist Club at the University campus on 15th June, 1966', *media release*, MC.JUN.22/66(PM), retrieved 20 June 2020, https://www.nas.gov.sg/archivesonline/data/pdfdoc/lky19660615.pdf

Turnbull, M 2017 National Security Statement, Transcript, 13 June 2017, retrieved 12 August 2020, https://www.malcolmturnbull.com.au/media/national-security-statement-tuesday-13-june-2017

Turnbull, M 2017 A Strong and Secure Australia, transcript, 18 July 2017, retrieved 12 August 2020, https://www.malcolmturnbull.com.au/media/a-strong-and-secure-australia

Newspaper articles/Blogs/Online commentary

Australian Associated Press 2019 'Vic Cops in Counter-terror Training Drill', *Canberra Times*, 12 October 2019, at https://www.canberratimes.com.au/ story/6434626/ vic-cops-in-counter-terror-training-drill/?cs=14231 (retrieved 12 February 2020)

ABC News 2020 'Bushfire response to be boosted by deployment of 3,000 ADF reservists, Prime Minister announces', *ABC News online*, 4 January 2020, retrieved 1 June 2023 https://www.abc.net.au/news/2020-01-04/australia-defence-reservists-to-help-in-bushfire-recovery/11840764

Ainge Roy, E 2019 "'I'm being watched': Anne-Marie Brady, the China critic living in fear of Beijing', *The Guardian*, 23 January, retrieved 2 June 2020, https://www.theguardian.com/world/2019/jan/23/im-being-watched-anne-marie-brady-the-china-critic-living-in-fear-of-beijing

Armin, J 2008 'Russian Cyberwar on Georgia', 9 October, retrieved 18 May 2020, http://hostexploit.com/downloads/view.download/4/9.html

Australian Associated Press 2019 'Asio investigating Chinese plot to plant spy in Australia's parliament after Liberal party member found dead', 25 November, retrieved 3 June 2020, https://www.theguardian.com/australia-news/2019/nov/25/asio-investigating-chinese-plot-to-plant-spy-in-australias-parliament-after-liberal-member-found-dead

Barrett, J 2019 'Solomon Islands government says Chinese lease of island unlawful', *The Sydney Morning Herald,* 26 October, retrieved 3 July 2020, https://www.smh.com.au/world/oceania/solomon-islands-government-says-chinese-lease-of-island-unlawful-20191026-p534gv.html

Bashen, Y 2020 'ASIO chases agents in the House', *The Weekend Australian*, 27–28 June

Baxendale, R 2020 'China-linked staffer's corona conspiracy', *The Australian,* The Nation, 2 June

——— 2020, 'Andrews staffer did Chinese propaganda course', *The Australian*, 29 June

Callick, R 2021 'Farewell to China's Most Sparkling Gem', Inquirer, *The Weekend Australian,* 20–21 January

Benson, S 2020 'Scott Morrison shoulders arms to China in 10 year $270bn plan', 30 June, *The Australian*, retrieved 1 July 2020, https://www.theaustralian.com.au/nation/politics/pm-shoulders-arms-to-china-in-10year-270bn-plan/news-story/1d130db628bde59abd6a02726bb94327

Birnes, WJ 2017 Foreword, in Liang, Q & Xiangsui, W 1999, *Unrestricted Warfare: China's Master Plan to Destroy America*. 2017 Kindle Edition, Shadow Lawn Press

Blainey, G 2020 'As the Pacific theatre opened, the nation was ill-prepared', *The Weekend Australian*, Inquirer, 15–16 August

Buchanan, E 2019 'Hybrid warfare: Australia's (not so) new normal', *The Strategist*, Australian Strategic Policy Institute, Canberra

Buckley, C 2013 'China takes aim at Western ideas', 19 August, *The New York Times*, retrieved 4 June 2020, https://www.nytimes.com/2013/08/20/world/asia/chinas-new-leadership-takes-hard-line-in-secret-memo.html?mcubz=1

Campbell, A 2019 'Political Warfare', ASPI Keynote Address, 18 June 2019 retrieved 20 July 2020 at https://www.youtube.com/watch?v=P7O40S9W7ks

Carr, B 2018, 'Bob Carr replies to China critics', *The Australian Financial Review*, 11 November, retrieved 11 June 2020, https://www.afr.com/opinion/despite-the-cold-warriors-australia-reverts-to-pragmatism-on-china-20181111-h17r92

Condon, M 2020, The boy who kicked the hornet's nest', *The Weekend Australian Magazine*, 30–31 May, retrieved 4 June 2020, https://www.theaustralian.com.au/weekend-australian-magazine/how-drew-pavlous-university-of-queensland-protest-enraged-china-and-started-a-free-speech-battle/news-story/82f5fd86413844c724e64322b11abb69

Croucher, G & Powell D 2019 'Our ties to China must be subtle and nuanced', *The Australian*, 30 October, retrieved 13 June 2020, https://www.theaustralian.com.au/higher-education/our-ties-to-china-must-be-subtle-and-nuanced/news-story/f1f75abe01896bc61d48633c12a24b82

Downer, A 2003, 'Neighbours cannot be recolonised', *The Australian*, 8 January 2003

Dupont, A 2019 'A New Type of War at Our Door', *The Australian*, Inquirer, 10 August, retrieved 5 February 2019, https://www.theaustralian.com.au/inquirer/a-new-type-of-war-at-our-door/news-story/243b8fcaee5e0fa8bcb817f7ce971073

Dupont, A 2020 'Australia must stand strong against Beijing's political warfare', *The Australian*, Inquirer, 4 January, retrieved 7 June 2020, https://www.theaustralian.com.au/inquirer/australia-must-stand-strong-against-beijings-political-warfare/news-story/1d88706d02cc0a8f65c24ff8d23540a2

Dupont, A 2019 'A New Type of War at Our Door', *The Australian*, 10 August, retrieved 05 February 2019 https://www.theaustralian.com.au/inquirer/a-new-type-of-war-at-our-door/news-story/243b8fcaee5e0fa8bcb817f7ce971073

Dupont, A 2020 'Who's afraid of the big bad wolf warriors', 25 July, *The Australian*, Inquirer, retrieved on 25 July 20202, https://www.theaustralian.com.au/inquirer/whos-afraid-of-chinas-big-badwolf-warriors/news-story/78ecb0971212799b83ea18ce6b10a12d

——— 2021 'China strategy: Get a bigger stick with which to protect ourselves', 30 April, The Australian, retrieved 30 April 2021 https://www.theaustralian.com.au/commentary/china-strategy-get-a-bigger-stick-with-which-to-protect-ourselves/news-story/9bf2fe8f2024ef6d61e0cc9e2df4eaf6

Dziedzic, S 2021 'New Zealand Trade Minister advises Australia to show China more 'respect', *ABC News*, 28 January, retrieved 17 February 2021, https://www.abc.net.au/news/2021-01-28/nz-trade-minister-advises-australia-to-show-china-more-respect/13098674

Fitzgerald, J 2018 'Australia on its own when managing foreign influence on Australian soil', *The Australian Financial Review*, 15 March, retrieved on 8 June

2020, https://www.afr.com/world/asia/australia-is-on-its-own-as-beijing-demonstrates-its-power-in-the-region-20180312-h0xbze

Freedberg Jnr, SJ 2018 'Electronic Warfare Trumps Cyber for Deterring Russia', Breaking Defense, 01 February, retrieved 20 October 2020, https://breakingdefense.com/

Galbally, F 2021 'We must bolster our cyber war defences', *The Australian*, 12 January, retrieved 12 January 2021

Grenfell, O 2019 Australian media's 'Chinese spy defection story' unravels', International Committee of the Fourth International, World Socialist Website, 5 December, retrieved 3 June 2020, https://www.wsws.org/en/articles/2019/12/05/wang-d05.html

Hamilton, C 2019 'Why Gladys Liu must answer to parliament about alleged links to the Chinese government', *The Conversation,* 11 September, retrieved 7 June 2020, https://theconversation.com/why-gladys-liu-must-answer-to-parliament-about-alleged-links-to-the-chinese-government-123339

Hamilton, C 2020 'Sheets pulled back in search for reds in bed with ALP', *Th Weekend Australian*, 27–28 June

Harris, R 2019 Gladys Liu's Beijing confession deepens dispute over loyalty, *The Sydney Morning Herald*, 11 September, retrieved 7 June 2020, https://www.smh.com.au/politics/federal/gladys-liu-s-beijing-confession-deepens-dispute-over-loyalty-20190911-p52qec.html

Jennings, P 2019 'The many ways China is pushing us around ... without resistance', 8 June, The Australian, retrieved 8 June 2020, https://www.theaustralian.com.au/inquirer/the-many-ways-in-which-china-is-pushing-us-around-without-resistance/news-story/2de4e239607062fd8859f486fd82403c

Jennings, P 2020 'China will be surprised how long it took us to act', *The Australian*, 8 June, p. 10, retrieved 8 June 2020, https://www.aspi.org.au/opinion/china-will-be-surprised-how-long-it-took-us-act-foreign-investment-laws

Keating, P 2019 'Video: "China is a great state' – Ex-Aussie PM Paul Keating tears into spooks over Beijing suspicions', *Hong Kong Free Press*, retrieved 13 June 2020, https://hongkongfp.com/2019/05/06/video-china-great-state-ex-aussie-pm-paul-keating-tears-spooks-beijing-suspicions/

Kearsley, J, Bagshaw, E & Galloway, A 2020 'If you make China the enemy, China will be the enemy': Beijing's fresh threat to Australia, 18 November, The Sydney Morning Herald, retrieved 4 February 2021, https://www.smh.com.au/world/asia/if-you-make-china-the-enemy-china-will-be-the-enemy-beijing-s-fresh-threat-to-australia-20201118-p56fqs.html

Kelly, J & Benson, S 2020 'Raided Labor MP in fundraising boast', *The Weekend Australian*, 27–28 June

Kelly, J & Packham, B 2020 'Sweeps first test of laws on foreign interference', *The Weekend Australian*, 27–28 June

Kelly, P 2020 'Shift on Beijing was Turnbull's gift to Morrison', *The Australian*, 29 April

Kennedy CM & Erickson, A 2016 'China's Uniformed, Navy-trained Fishing 'Militia'', CIMSEC series, part 2, 17 May, retrieved 15 June 2020, https://www.maritime-executive.com/editorials/chinas-uniformed-navy-trained-maritime-militia and http://cimsec.org/new-cimsec-series-on-irregular-forces-at-sea-not-merely-fishermen-shedding-light-on-chinas-maritime-militia/19624

Lindley-French, J 2019 Briefing: Complex Strategic Coercion and Russian Military Modernisation, *Policy Perspective*, Canadian Global Affairs Institute, Calgary, January, retrieved 21 October 2020, https://d3n8a8pro7vhmx.cloudfront.net/cdfai/pages/4117/attachments/original/1548354852/Complex_Strategic_Coercion_and_Russian_Military_Modernization.pdf?1548354852=

The Lindley-French Analysis: Speaking Truth Unto Power, 2019 blog, 9 January, retrieved 21 October 2020, http://lindleyfrench.blogspot.com/2019/01/briefing-complex-strategic-coercion-and.html

McKenzie, N, Sakkal, P & Tobin, G 2019 'China tried to plant its candidate in Federal Parliament, authorities believe', *The Age*, 24 November, retrieved 8 June 2020, https://www.theage.com.au/national/china-tried-to-plant-its-candidate-in-federal-parliament-authorities-believe-20191122-p53d9x.html

Korporaal, G 2021 Hong Kong activists arrested in China show of force, *The Australian*, 7 January 2021, retrieved on 7 January 2021, https://www.theaustralian.com.au/world/hong-kong-activists-arrested-in-china-show-of-force/news-story/62299971397a5543b05fb1f6b8a504a9?utm_source=&utm_medium=&utm_campaign=&utm_content=

McGuinness, D 2017 How a cyberattack transformed Estonia, BBC News, 27 April 2017, retrieved 16 December 2020, https://www.bbc.com/news/39655415

Medcalf, R 2018 'Silent Invasion: the question of race', *The Interpreter*, The Lowy Institute, 21 March, retrieved 13 June 2020, https://www.lowyinstitute.org/the-interpreter/silent-invasion-question-race

Milhiet, P 2017 'China's Ambition in the Pacific: Worldwide Geopolitical Issues', Institute de Relations Internationales et Strategiques, Asia Focus #49 – Asia Program, retrieved 20 June 2020, https://www.iris-france.org/wp-content/uploads/2017/10/Asia-Focus-49.pdf

Moses, A 2008 Georgian websites forced offline in 'cyber war', 12 August, *The Sydney Morning Herald*, retrieved 4 February 2021, https://www.smh.com.au/technology/georgian-websites-forced-offline-in-cyber-war-20080812-gdsqac.html

Munro, K 2019 'Australian attitudes to China shift: 2019 Lowy Poll', 27 June, Lowy Institute, retrieved 21 June 2020, https://www.lowyinstitute.org/the-interpreter/australian-attitudes-china-shift-2019-lowy-poll

Needham, K 2019 'Beijing supports 'patriotic' protests against Hong Kong students in Australia', *The Sydney Morning Herald*, Fairfax Media, Sydney, retrieved 10 October

2019, https://www.smh.com.au/world/asia/beijing-supports-patriotic-protests-against-hong-kong-students-in-australia-20190820-p52iu7.html

Norington, B 2020 'MP China-link raids underline ASIO's anxiety', *The Australian*, 29 June

Packham, B 2020 'Influencer cosied up to Chinese leaders', *The Australian,* The Nation, 1 June

Oliver, A & Kassam, N 2020 'Happy to hop away from these bounders: The 2020 Lowy Institute Poll finds us squaring up to an uncertain future', *The Australian*, Commentary, 24 June 2020

Packham, B 2020 Cyber spy agency on high alert over hack, *The Australian*, 24 December, retrieved 24 December 2020, https://www.theaustralian.com.au/nation/cyber-spy-agency-on-high-alert-over-hack/news-story/a4879aac7be8536b662af8b29f2d3d20?utm_source=&utm_medium=&utm_campaign=&utm_content=

Packham, B 2021 China dangles $39bn carrot to build city on our doorstep, *The Australian*, 5 February, retrieved 5 February 2021 https://www.theaustralian.com.au/nation/china-dangles-39bn-carrot-to-build-city-on-our-doorstep/news-story/0c8b1219fb3db6d51a5a8d95e224f50b?utm_source=&utm_medium=&utm_campaign=&utm_content=

Paterson, T 2019 'The Grey Zone: Political Warfare is Back', The Interpreter, The Lowy Institute, retrieved 19 May 2020 https://www.lowyinstitute.org/the-interpreter/grey-zone-political-warfare-back

Riordan, P 2018 'China to host Pacific Islands meeting ahead of APEC', *The Weekend Australian*, 10 July, retrieved 20 June 2020, https://www.theaustralian.com.au/national-affairs/foreign-affairs/china-to-host-pacific-islands-meeting-ahead-of-apec/news-story/961c0cb7fe2ab07e5fdf7166eb7fa005

Rudd, K 2024 'Short of War: China's gray zone strategies are gathering in intensity, *The Washington Post*, 6 June, retrieved 23 June 2024, https://www.washingtonpost.com/opinions/2024/06/06/kevin-rudd-china-taiwan-deterrence/

Sheridan, G 2020 Scott Morrison right to say times as dangerous as 1930s, *The Australian*, 2 July, retrieved 26 July 2020, https://www.theaustralian.com.au/commentary/scott-morrison-backs-tough-words-with-robust-actions/news-story/ce0f4f5ffecc97e90ab415661c23cbbc and https://www.theaustralian.com.au/author/Greg+Sheridan

Slattery, L 2020 'Hitting Home', 25–26 April, *The Weekend Australian*, retrieved 4 June 20202, https://www.theaustralian.com.au/weekend-australian-magazine/swimmer-mack-hortons-family-reveals-fallout-from-drug-protest/news-story/a3f11ec2851c90b4171b8021c168a200

Sun, W 2019 'What do we learn about the experience of Mandarin-speaking migrants from Chinese-language media in Australia?', *ABC Religion and Ethics*, 25 November 2019, retrieved 14 June, https://www.abc.net.au/religion/what-we-learn-from-chinese-language-media-in-australia/11735478

Sweeny, L et al. 2017 'Sam Dastyari resignation: How we got there', 12 December, *ABC News*, retrieved 3 June 2020, https://www.abc.net.au/news/2017-12-12/sam-dastyari-resigns-from-parliament/9247390 and https://www.abc.net.au/news/2017-12-12/sam-dastyari-resignation-how-did-we-get-here/9249380

Sakkal, P & Mackenzie, N 2019 'How Nick Zhao made enemies, faced charges, and was allegedly asked to spy for China', *The Age*, 29 November, retrieved 8 June 2020, https://www.theage.com.au/national/how-nick-zhao-made-enemies-faced-charges-and-was-allegedly-asked-to-spy-for-china-20191128-p53ezs.html

Tillett, A 2021 'Chinese city plan on PNG island highlights concerns about Beijing', *Australian Financial Review*, 5 February, retrieved 9 February 2021, https://www.afr.com/politics/federal/chinese-city-plan-on-png-island-highlights-concerns-about-beijing-20210205-p56zwm

Varga, R 2020 'Key BRI advisor linked to Communist Party', *The Australian,* The Nation, 1 June

Varghese, P 2020 'How to best manage our relationship with Beijing: Adopting a policy of engaging and constraining China suits Australia's interest far better, *The Weekend Australian,* 27–28 June 2020

Walden, M 2021 China again blames Australia for diplomatic spat, issues warning to Five Eyes intelligence partners, *ABC News*, 20 November, retrieved 17 February 2021, https://www.abc.net.au/news/2020-11-20/china-blames-confrontational-australian-government-for-spat/12902774

Wall, J, 2020 China to build $200 million fishery project on Australia's doorstep, *The Strategist*, Australian Security Policy Institute, 8 December, retrieved 9 February 2021, https://www.aspistrategist.org.au/china-to-build-200-million-fishery-project-on-australias-doorstep/

White, H 2017 'China's Power and the Future of Australia', *ANU TV*, YouTube, retrieved 13 June 2020, https://www.youtube.com/watch?v=8JWLnaVvuJg&feature=youtu.be

——— 2017 'We need to talk about China', Asia and Pacific Policy Society, 4 May, retrieved 13 June 2020, https://www.youtube.com/watch?v=8JWLnaVvuJg&feature=youtu.be

White, N 2020 'Why two of Australia's richest men are backing China in diplomatic row sparked by push for a coronavirus inquiry as Beijing threatens to destroy our economy as revenge,' *The Daily Mail – Australia*, 4 May, retrieved 3 July 2020, https://www.dailymail.co.uk/news/article-8272347/Andrew-Twiggy-Forrest-Kerry-Stokes-China-diplomatic-row.html

Zhou, N & Smee B 2020 'We cannot be seen: the fallout from the University of Queensland's Hong Kong protests', *The Guardian*, 4 August, retrieved 8 June 2020, https://www.theguardian.com/australia-news/2019/aug/04/we-cannot-be-seen-the-fallout-from-the-university-of-queenslands-hong-kong-protests

Podcasts/Documentaries

Fox News 2020 'The Strike on Soleimani: Exclusive Photos of Special Op's Bomb Damage Assessment', YouTube, 10 January, retrieved 14 January 2020, https://www.youtube.com/watch?v=Hue99QYu3kk

Jennings, P 2016 Peter Jennings on China, Australia and soft power', podcast, 20 September, ASPI, retrieved 3 June 2020 https://www.aspi.org.au/video/peter-jennings-china-australia-and-soft-power

McKenzie, N 2019 'China's Spy Secrets', *60 Minutes*, 9 Entertainment Co, retrieved 3 June 2020, https://9now.nine.com.au/60-minutes/chinas-spy-secrets/9f3b7622-28a1-4823-888f-878aa2146698

Langford, I, 2017 podcast, Covert Contact podcast, 25 September 2017, *Australian Special Forces*, retrieved 10 November 2017, http://covertcontact.com/tag/colian-langford/

Langford I 2021 Presentation to the 2021 Defence Reserves Association Conference, 14 August 2021, retrieved 03 April 2023, https://www.youtube.com/watch?v=SeV3XipWgQw&list=PLVj2rO_2uYciJjUk9MNbNUk4sI0LbUBIL&index=10

Matisek, J & Bertram, I 2017 'The Death of American Conventional Warfare: It's the Political Willpower, Stupid', *Strategy Bridge* podcast, 5 November, p. 4, retrieved 11 November 2017, https://thestrategybridge.org/the-bridge/2017/11/5/the-death-of-american-conventional-warfare-its-the-political-willpower-stupid

McKenzie, N 2019 'World Exclusive: Chinese spy spills secrets to expose Communist espionage', *60 Minutes* 24 November 2019 https://www.youtube.com/watch?v=zdR-I35Ladk

Mulyanto, R & Tobin, M 2019 'East Timor's China friendship won't compromise its national interests: foreign minister', *This Week in Asia*, 23 August, retrieved 20 June 2020, https://www.scmp.com/week-asia/politics/article/3023934/east-timors-china-friendship-wont-compromise-its-national

Index

www.ingramcontent.com/pod-product-compliance
Lightning Source LLC
Chambersburg PA
CBHW050841270326
41930CB00019B/3423